Contemporary Geopolitical Alignment in the Global South

Mukunda Mishra · Andrews José de Lucena ·
Suresh Kumar · Siddhartha Sankar Manna
Editors

Contemporary Geopolitical Alignment in the Global South

Security, Peace and Sustainable Development

Springer

Editors
Mukunda Mishra
Department of Geography
Dr. Meghnad Saha College
Uttar Dinajpur, West Bengal, India

Andrews José de Lucena
Department of Geography
Federal Rural University of Rio de Janeiro
Rio de Janeiro, Brazil

Suresh Kumar
Department of African Studies
University of Delhi
New Delhi, Delhi, India

Siddhartha Sankar Manna
Department of Political Science
University of Gour Banga
Malda, West Bengal, India

ISBN 978-3-031-81976-6 ISBN 978-3-031-81977-3 (eBook)
https://doi.org/10.1007/978-3-031-81977-3

This Springer imprint is published by the registered company Springer Nature Switzerland AG
The registered company address is: Gewerbestrasse 11, 6330 Cham, Switzerland

Foreword

It is my great privilege to introduce inquisitive readers to this timely and pertinent volume, edited by a team of social scientists who seek to find out the dynamic equation of geopolitics as envisaged in the global south amidst the acute crisis global set-up, particularly, the human-induced climate change and long-run effects of the worst pandemic of the millennium. The book entitled *Contemporary Geopolitical Alignment in the Global South: Security, Peace and Sustainable Development* accommodates a bunch of curated chapters, duly edited by Mukunda Mishra, Andrews José de Lucena, Suresh Kumar, and Siddhartha Sankar Manna, who have been critically writing on the social, economic, and political actors of development in the Global South for a long time.

South and Southeast Asia is an evolving and intricate tapestry of power, security, and conflicts. The geopolitical significance of the Indian Ocean, India's strategic position, and emerging post-cold-war regional political equations shape the region's future, where understanding these dynamics is essential to navigate the complexities and work toward ensuring stability, cooperation, and prosperity in this region. Both the land and sea are geopolitically significant here. While the South Asian landmass, nestled between the Arabian Sea and the Bay of Bengal, is accredited with its strategic location, the abundant natural and cultural resources encompass its attribute to be the focal point of global powers. Moreover, the region, being the home to several nuclear powers, the evolving territorial disputes and historical conflicts that often mold the geopolitical equations of South Asia, need careful resolution through regional cooperation and integrity for the sake of building peace and sustainability. There is often some sort of exercise by external powers in the region which also acts as the catalyst for South Asia's geopolitical landscape. For the purpose of regional security and global stability, understanding the instances of South Asian geopolitics is essential.

This book interestingly presents the regional geopolitical processes in the global south within a single broad frame, which has been rather a necessity than an obligation at this current time. Geopolitics is no longer that simple and monotonous as well.

While, until the 1990s, concerning the great power competition, European policymakers viewed Asian, African, and Latin American nations as easily "pliable," and largely insignificant where these states were in the policy papers as the sources of raw material, "populous" markets, and investment destinations, their voices, priorities, and concerns were mostly disregarded in the international platforms that were often crafted by the West. In a changing scenario, the global south has been re-emerging with its new identities; they are climbing up the international agenda which is important to be documented which this book certainly seeks to do.

The book, with its wide coverage, promises to be of great usefulness to serious students, teachers, and researchers who are interested in the broader dimension of geopolitics plying throughout the global south which is, observed, researched, and analyzed by the scholars of the global south themselves, searching pathways for human society for a safe, peaceful and collaborating future toward one earth one nation.

January 2026

Pankaj Kumar, Ph.D.
Assistant Secretary-General
International Geographical
Union (IGU), & Professor
Department of Geography
Delhi School of Economics
University of Delhi
New Delhi, India

Preface

A century-long journey from Ratzel's organic state theory to social Darwinism to further evolved through the conceptualization of geopolitics that sets to address the political power linked to geographic space, which ultimately founded the cornerstone of critical geopolitics as the product of the post-structural approach. Accentuating the balanced usage of territorial parts of the world to use the biological natural resources within boundaries for safeguarding those properties accessible for usage by the states/independent states reinforces the notion of geo-politics since the 1960s. Thus, the concept of geopolitics becomes significant in the field of Political Geography and Earth's Geography (human and physical), influencing international relations, specifically in political, social, economic, defense, and cultural ones. A sovereign state geopolitically emphasizes de jure status and establishes a relationship with other states on parity, but de facto states with partial international recognition (for example, Western Sahara) have geopolitical significance. Along with it, the de jure States having sub-national geopolitical entities either choose federal, quasi-federal, federation, or confederation structure of their governance get partial recognition internationally. In a theoretical sense, geopolitical understanding uses traditional approaches to align their role with great powers, but globalization has changed the political and ideological understanding of states vis a great power. Therefore, geographical and terrestrial variables of the Foreign Policy of a State use Geopolitical understanding to initiate intercontinental political behavior and act accordingly to maintain international relations. Today, these variables encompass regional studies, micro and macro-climate, landscape, human ecology, population studies, natural resources, and applied science of the assessed region.

Global South, having one or other kind of variables, acts as a living laboratory in which Geopolitical India, as the largest democracy in the world, establishes its sustainable credentials. Geopolitical allocation of diversities within a federation requires not always to strictly follow the boundary lines of the component units of the state. Geopolitically, African, Asian, and Latin American countries have experienced the re-drawing of their national borders at the time of independence and, in the course of their growth toward maturity, do create union through the real functions of the federal government and the component units of the state. Global South had a dictatorship, military rule, and a one-party system to a multi-party system and are shaping the sustainable development mechanism in their own style. Despite its Geopolitical significance (it acts as a bridge and connects South Asia with Central Asia), Afghanistan faces a current crisis of political instability and a state without an elected government. The question—whether democracy is a necessary element for economic development or an obstacle because of massive development needs—storms the intellectual debates. Understanding the link between democracy and the development of a country needs explicit crystal knowledge about the nature of the country's economic development, its socio-political history, its culture of treating natural resources, the form of governance, and the role of people in the development process.

The pertinent discussion is storming the scholarly minds that the geopolitical landscape is shifting sharply; however, the future is unpredictable amidst the transition point where all equations are not yet clear. While some economic consequences are coming into the light of the day, the long-range social changes are still in the dark. The United States of America (USA), which played historically as the enforcer of world geopolitical order, has concentrated on internal policy issues rather than on foreign policies. Perhaps the absence of the US from the scene has allowed the rise of other global powers, e.g., China and Russia, who have been competing for the role of a global leader with the USA. Amidst these, the European Union (EU) traces an R&D landscape with collaborative industrial action in the defense and aerospace sectors and the support of the financial tool of the European Defense Fund (EDF).

Literature on China's geo-economic propaganda in almost every corner of Africa is ample and vivid; however, the authors are not in common agreement on whether China's operations in Africa are rather favorable, with infrastructure being expanded and modernized. Some critical thinkers speculate it is the long-term strategic objective to create a new world order favorable to China's authoritarian brand of politics or its "socialist market economy." On the other hand, there are arguments favoring that loans and aid from China have an underlying agenda: to make African states dependent on China and make them more compliant in international contexts, where China is dependent on the voices of these states, such as within the UN, or in many sensitive issues, e.g., Taiwan and Hong Kong matters.

The dramatic global events unleashed by Russia's large-scale invasion of Ukraine are increasing the demand for U.S. "nuclear umbrellas" in Asia and Europe; South Korea and Japan are debating the redeployment of U.S. nuclear weapons; Finland and Sweden have applied to join NATO. However, above all, the tendency for future expansion of the U.S. nuclear arsenal is potentially reversing a three-decades-long trend of arms reductions and creating new dangers of nuclear usage. The security of sovereignty throughout the world is possibly getting threatened.

However, these fast-moving global trends must be understood and explained through 'non-military' variables that could explore the shake of the tremor reaching far Africa. For quite a few decades of changing geopolitical equations, the non-military factors have appeared more relevant than ever before. The world is witnessing the applications of "geo-economics" to advance geopolitical agendas through economic tools, e.g., trade control, investment policy, economic and financial sanctions, energy and commodities, aid policies, data management, communication infrastructures, cyberspace control, and many more. An outburst of war at any node of this network would essentially be carried forward to other nodes not essentially as a war, but a range of social and economic change, at which the poor suffer most.

The post-COVID-19 social and economic transformation has been essentially playing a role in the internal policies and governance of the global south states. For example, Sri Lanka, a country of 22 million people, is caught in its worst financial crisis in seven decades after its foreign exchange reserves went to record lows while the dollars are running out to pay for essential imports, including food, medicine, and fuel. The situation acts as the catalyst for such states to set closer associations with the geo-economic network. As a result, it makes the global south states more vulnerable to receiving the shock of an outburst happening at any farthest node in the network.

"Climate change" poses another crucial driving force of geopolitics, and its role has changed markedly in the last decade. From being a taker of the prevailing geopolitical winds, driven by other issues like security and trade, Climate has shifted to be a shaper of world geopolitics. Amidst the cycles of the COVID-19 pandemic, 2020 has become the second warmest year, with record-high key greenhouse gas concentrations in the atmosphere and evidently high surface temperatures accompanied by unprecedented numbers of climate-related weather disasters like cyclones, droughts, and vast wildfires. East African states, i.e., parts of Somalia, Djibouti, Ethiopia, and Kenya, are experiencing the driest conditions and hottest temperatures ever, keeping as many as 13 million people under acute food and water shortages during mid-2021, and a projected 25 million will face a similar fate by mid-2022. Climate change-induced crisis becomes widespread in that part of the world that is least able to cope. Africa as a whole contributes to only about two to three percent of global emissions that cause climate change, but this continent suffers the worst. However, the Chinese President pledged at the United Nations General Assembly in September 2020 that China will be reaching the peak of carbon dioxide emissions by 2030, and after that, cutting the net greenhouse gas emissions to zero to achieve carbon neutrality by 2060. Sincere cooperation and prioritization of climate agendas by the state leaders at every corner of the world could not be achieved through dialogues but through proper action.

With this backdrop, the editors have carefully chosen the chapters that provide a framework for global relations by linking the geostrategic practices, governance, and sustainable development to address the contemporary debates at the crossroads of geographical space and international politics from several disciplinary and methodological standpoints. These chapters will speak about whether the geopolitical alignment could abreast the physical arrangements and sustainability; what the necessities of developing a refined understanding of the topographical and multiscale dynamics of international politics; and why it needs concrete nature of actions, resilient and attention, particularly the abounding Global South states.

Itahar, West Bengal, India
Rio de Jeneiro, Brazil
New Delhi, India
Malda, West Bengal, India
November 2024

Mukunda Mishra
Andrews José de Lucena
Suresh Kumar
Siddhartha Sankar Manna

Acknowledgments

We spent almost more than two years to finalize this book. We must thank all of our friends and families, colleagues, students, and all others connected with this venture would not have been completed without whose advice, mental support, and inspiration to write.

We express our most profound sense of indebtedness to Prof. Pankaj Kumar, Department of Geography, University of Delhi, and Assistant Secretary-General, International Geographical Union (IGU). We thank him for his kind advice and valuable time writing the 'Foreword' for this volume.

Many scholarly minds behind the scenes have done commendable works of rigorous reviewing, and making valuable advice on the manuscripts for the volume which helped substantially maintain the quality of articles. The editors are thankful to all the reviewers of the chapters.

In an edited volume, the role of the contributing authors is precious. We convey our sincere thanks to all contributors for offering us the opportunity to include their valuable works in this volume.

Constructive editorial advice and constant support from Zachary Romano, *the* Senior Publishing Editor, at Springer New York, remains unparalleled. We acknowledge the support of the entire team of Springer Nature associated with the publishing process, disseminating their respective roles as perfectly as ever.

Itahar, West Bengal, India — Mukunda Mishra
Rio de Jeneiro, Brazil — Andrews José de Lucena
New Delhi, India — Suresh Kumar
Malda, West Bengal, India — Siddhartha Sankar Manna
November 2024

Contents

Editors and Contributors

About the Editors

Mukunda Mishra is an associate professor, Department of Geography, and designated Vice Principal of Dr. Meghnad Saha College in West Bengal, India. Dr. Mishra completed his postgraduate studies in Geography and Environmental Management at Vidyasagar University (receiving top rank in both the B.Sc. and M.Sc. panels of merit) and held a Ph.D. in Geography from the same University. He was selected for the prestigious National Merit Scholarship by the Ministry of Human Resource Development, Government of India. His research chiefly focuses on analyzing unequal human development, regional planning, and multi-criteria predictive model building. He has more than twelve years of hands-on experience in dealing with development issues at the ground level in various districts of eastern India. Dr. Mishra has in his credit to publish one monograph and nine edited research volumes so far from the house of Springer-Nature. Dr. Mishra has more than thirty research articles and book chapters published in the journals and books of international repute. He serves as the reviewer of several reputed journals published by Springer-Nature, Elsevier and other publishing house of international repute.

Andrews José de Lucena is an associate professor, Federal Rural University of Rio de Janeiro, Brazil. He has completed his Master in Geography from the State University of Rio de Janeiro and holds a Ph.D. in Atmospheric Sciences from the Alberto Luiz Coimbra Postgraduate Institute and Engineering Research (COPPE/UFRJ). His research interest is chiefly concerned with the urban climate and its research methods, climate change, and environment. He participated and integrated research projects with several professionals and institutions in the area of Climatology and Urban Meteorology, Atmospheric-Hydrological Modeling, Remote Sensing in Urban Areas, and Urban Development in the Metropolitan Region of Rio de Janeiro. He coordinates the Integrated Laboratory of Applied Physical Geography (LIGA/UFRRJ) and integrates the Environmental Satellite Applications Laboratory of the Meteorology Department (LASA/UFRJ). He manages the website www.climatologia.com.br with information about land surface temperature (LST) of the Metropolitan Area of Rio de Janeiro from 1984 to the present.

Suresh Kumar is a professor in the Department of African Studies at the University of Delhi. He shouldered the responsibilities as Head of Department and Coordinator of Center for African Studies, University Grants Commission (UGC) Area Study Program for 2015–2019 and 2009–2012. He is a chief editor of Africaindia.org (http://www.africaindia.org). Professor Kumar is a passionate traveler who has visited forty countries for research and teaching assignments. He is Chief Collaborator of the UGC Major Research Project on Geopolitics Federalism: A Vision of North and South Sudan, 2011, University research projects on Geological Evidences and Social Connectivity of India and Africa, 2015, Geopolitics and the Issue of Western Sahara, 2013, Geopolitics and the Issue of Darfur in Sudan, 2012, Political Economy and Social Issues in Eritrea, 2009 and Democracy and social Issues in South Africa and India, 2003. Recently, he finished International Research Project on Religion, Yoga, and Education in India and South Africa sponsored by Indian Council of Social Science Research-South Africa National Institute for the Humanities and Social Sciences, 2019 and completed Hashim Mbita Research Project on India, Indonesia, Sri Lanka and Yugoslavia (NAM) and Liberation Struggle of Southern Africa sponsored by Southern African Development Community (SADC), Tanzania, 2010. He is currently engaged with UNESCO Research Group on African and Brazilian knowledge, practices, education and histories in a Southern Perspective, Porto University, Brazil, 2012–Continue. He articulated his opinion on different aspects of Africa before the United Nations Organization (UNO, 2012), International Geographical Congress (IGC, IGU 2004 and 2008), Ronald H. Brown Institute (2008), World Policy Council (WPC 2014), European Commission of African Studies (ECAS), Council for the Development of Social Science Research in Africa (CODESRIA 2015), IBSA, BRICS-IKS (2016, 2018 and 2019), and many others. He has completed LLB successfully out of his sheer interests from University of Delhi, India in 2016.

Siddhartha Sankar Manna is an assistant professor in the Department of Political Science at the University of Gour Banga. He was awarded with the Ph.D. from the Vidyasagar University. The focal theme of his research is the analysis of international relations, political theory, democratic politics, and human development. He published more than twenty papers in reputed national and international journals.

Contributors

Thiago Jeremias Baptista Department of Human Geography, Federal Rural University of Rio de Janeiro, Rio de Janeiro, Brazil

Phillipe Valente Cardoso Department of Cartography and Geoprocessing, State University of Rio de Janeiro, Rio de Janeiro, Brazil

André Santos da Rocha Department of Geography, Federal Rural University of Rio de Janeiro (UFRRJ), Rio de Janeiro, Brazil

Vinicius da Silva Seabra Department of Cartography and Geoprocessing, State University of Rio de Janeiro, Rio de Janeiro, Brazil

Heitor Soares de Farias Department of Geography, Federal Rural University of Rio de Janeiro, Rio de Janeiro, Brazil

Leandro Dias de Oliveira Department of Geography at Federal Rural, University of Rio de Janeiro (UFRRJ), Rio de Janeiro, Brazil

Gustavo Mota de Sousa Department of Cartography and Remote Sensing, Federal Rural University of Rio de Janeiro, Rio de Janeiro, Brazil

Jonathan Christian Dias dos Santos Department of Geography, Federal Rural University of Rio de Janeiro (UFRRJ), Rio de Janeiro, Brazil

Somenath Halder Department of Geography, Kaliachak College, Malda, West Bengal, India

Pablo Ibañez Department of Geography, Federal Rural University of Rio de Janeiro (UFRRJ), Rio de Janeiro, Brazil

Suresh Kumar Department of African Studies, Faculty of Social Sciences, University of Delhi, Delhi, India;
Department of African Studies and Centre for African Studies, University of Delhi, Delhi, India

Siddhartha Sankar Manna Department of Political Science, University of Gour Banga, Malda, West Bengal, India

Wladimir Mejía Ayala Pedagogical and Technological University of Colombia, Tunja, Colombia

Mukunda Mishra Dr. Meghnad Saha College, Itahar, West Bengal, India

Asis Mistry Department of Political Science, University of Calcutta, Alipore, Kolkata, West Bengal, India

Sanjoy Saha Kaliachak College, Sultanganj, Malda, West Bengal, India

Patricia Teixeira Santos Department of History and Coordinator of Center for Studies On the History of Contemporary Africa (NEHAC), Sao Paulo Federal University (UNIFESP), Sao Paulo, Brazil

Marcos Aurelio Saquet State University of Western Paraná, Cascavel, Brasil

Júlio César Suzuki University of São Paulo, São Paulo, Brasil

Abbreviations

ACP	(Preferential trade relationship with the) African, Caribbean, and Pacific
ADB	Asian Development Bank
ARCA	ASEAN Regional Center for Arbitration
ARF	ASEAN Regional Forum
ARSA	Arakan Rohingya Salvation Army
ASEAN	Association of Southeast Asian Nations
AU	African Union
BRI	Belt and Road Initiative (by Chinese Government)
BRICS	(An intergovernmental organization comprising) Brazil, Russia, India, China, South Africa, Iran, Egypt, Ethiopia, and the United Arab Emirates
CCP	Climate Change Phenomena
CNRP	Cambodia National Rescue Party
CSR	Corporate Social Responsibility
DACE	District Advisory Committee on Environment (of Ladakh)
EEZ	Exclusive Economic Zone
EITI	Extractive Industry Transparency Initiative (in Africa)
FDI	Foreign Direct Investment
GCC	Global Climate Change
GDP	Gross Domestic Products
IBSA	India, Brazil, and South Africa (dialogue forum)
ICJ	International Court of Justice
IDP	Internally Displaced Person
INSTC	International North-South Transport Corridor
ITLOS	International Tribunal for the Law of the Sea
MCC	Millennium Challenge Corporation (of Nepal)
MDG	Millennium Development Goals
MERCOSUR	The Southern Common Market (Spanish abbreviation)
NATO	North Atlantic Treaty Organization
NGO	Non-Governmental Organization

OPIP	Organization of the Indigenous Peoples of Pastaza
PALOP	Países Africanos de Língua Oficial Portuguesa
PLAN	People's Liberation Army Navy (of the Republic of China)
PRC	People's Republic of China
ROC	Republic of China
SAARC	South Asian Association for Regional Cooperation
SACU	Southern African Customs Union
SDG	Sustainable Development Goals
SSC	South-South Cooperation
UHS	Unified Health System (in Brazil)
UN	United Nations
UNASUR	Union of South American Nations
UNCLOS	United Nations Convention on the Law of the Sea
UNESCO	United Nations Educational, Scientific and Cultural Organization
UT	Union Territory (of India)
XUAR	Xinjiang Uyghur Autonomous Region (in China)

List of Figures

List of Tables

Chapter 1
Boundary Disagreements and Micro-Conflicts in South and Southeast Asia: A Comprehensive Review

Siddhartha Sankar Manna

Abstract The regional conflict systems are often considered by their complexity. This effort has basically attempted to analyze the different types of issues in context of a comprehensive and holistic manner. The primary focus of this book goes beyond state and interaction of the state with the nationalities in South Asia. This chapter identifies the several complexities and various actors, reasons, and structural arrangements etc. Regional and territorial conflicts in Asia remain a challenge to the peace, stability, and prosperity of the region. This chapter focuses on the several ways for territorial issues: (i) Territorial disputes are about the land; (ii) Positional disputes and boundary line; (iii) Functional disputes.

Keywords Geopolitics · border · Ethno-linguistic · Religious identities · Regional organizations

1.1 Introduction

South and Southeast Asian countries continue to face difficulties due to micro-conflicts. While greater geopolitical tensions tend to eclipse these smaller wars, they still have far-reaching effects on the people who live in those regions. Micro-conflicts are discussed in depth, along with their origins, regional effects, and potential solutions, in this article. For a long time, South and Southeast Asian micro-conflicts have troubled the area and the rest of the world. When compared to big interstate wars, these conflicts may seem little, but they nonetheless pose a threat to regional peace and prosperity if they are allowed to escalate. Understanding the roots of these conflicts, gauging their influence on the region, and investigating potential methods of conflict resolution are all vital steps toward addressing and resolving them. South and Southeast Asia have long been plagued by micro-conflicts, or small-scale conflicts with a

S. S. Manna (✉)
Department of Political Science, University of Gour Banga, Malda, West Bengal, India
e-mail: manna.siddhartha@yahoo.in

M. Mishra et al. (eds.), *Contemporary Geopolitical Alignment in the Global South*,
https://doi.org/10.1007/978-3-031-81977-3_1

high potential for escalation. Ethnic and religious differences also play a significant role in these conflicts. South and Southeast Asia are home to diverse ethnic and religious communities, and the demarcation of borders has often led to the division of these communities. This has created a fertile ground for tensions and micro-conflicts to arise, as communities on either side of a border may feel marginalized or mistreated. Numerous causes, such as long-simmering resentment, racial tensions, territorial disputes, and unstable governments, contribute to the persistence of such wars. Facilitating regional peace and stability requires a focus on resolving these smaller issues. This article will investigate the origins of micro conflicts, the damage they do to a region, and possible solutions.

Unresolved colonial tensions, unclear or contested border demarcations, the historically porous nature of regions of borders, dispute over natural resources, and the "less-" or "un-governed" nature of a number of these locations are several variables that contribute to the identification of border regions as sites of latent conflict (Avis, 2020). Half of the nations in South and Southeast Asia are affected by subnational conflicts with strong trans-border features, making it the most common and longest-lasting form of conflict in the region. Asia's territorial disagreements continue to threaten Asia's geographic stability, development, and safety. Territorial disputes between states are the most prone to escalate into open conflicts. Conflicts over contested territory have been linked to a variety of factors, including political and economic interests, normative motivations, and competition over limited natural resources. Different territorial disputes have different impacts on the bilateral and multilateral relations of the governments concerned, and they might have different causes. Many different types of territorial disputes exist. There is a wide range of territorial conflicts throughout Asia, each with its own unique history, scope of area at issue, and impact on bilateral and multilateral ties between nations (Bukh, 2020). The three types of border disputes presented by Brunet-Jailly (2015) are: (i) territorial; (ii) positional; and (iii) functional. In accordance with this classification system:

- Land is at the centre of territorial conflicts. Most difficult because they threaten the very existence of nations;
- Positional disagreements occur when there is general agreement on a border but controversy about where the line of demarcation should be formed;
- Functional conflicts involve different perspectives on what role a given boundary should play rather than disagreements over land or the borderline itself.

Conflicts over scarce natural resources are a common feature of these classifications. Many actors, reasons, organisational circumstances, and dynamical processes contribute to the complicated nature of conflicts within regional systems. Understanding the regional patterns regarding violence is complicated by this level of intricacy (Parry et al., 2007). It is also not apparent what factors contribute to the emergence of a regional dispute system or how violence spreads across geographic areas. Who the major players are, how they interact, and how much they shape the conflict dynamics in every given border issue are all questions that remain unsolved. Boundary disagreements and micro-conflicts have long plagued the region of South and Southeast Asia, causing tensions and hampering regional stability and cooperation. These disputes have deep historical roots and arise from a variety of complex factors, including colonial legacies, territorial claims, and ethnic and religious differences.

1.2 Understanding the Root Causes of Micro-Conflicts

South and Southeast Asian micro-conflicts typically stem from a tangled web of historical, political, and socio-economic variables. Ethnic and religious tensions have been identified as contributing factors. Miscommunications and disputes are possible because of the region's diverse population, which comes from a variety of cultural and religious origins. Identity politics, in which ethnic, religious, or linguistic distinctions feed tensions and ultimately lead to bloodshed, is a major contributor to these conflicts. Micro-conflicts have emerged in the region because of the legacy of colonialism and the ensuing battle for independence (Yusuf, 1994). Understanding the origins of micro-conflicts is crucial for finding lasting solutions to them. Historical grudges, such as colonial legacies or unresolved conflicts, are a big contributor. Disagreement and hostility between people of different races and religions are common results of such injustices. Another notable dispute is the South China Sea conflict, involving multiple countries such as China, Vietnam, the Philippines, and Malaysia. The territorial claims and competing interests in this resource-rich area have led to increased militarization and strained diplomatic relations. Other significant boundary disputes in the region include the dispute between India and China over the Line of Actual Control, the ongoing conflict between Thailand and Cambodia over the Preah Vihear Temple, and the territorial disagreements between Myanmar and Bangladesh. These disputes highlight the complex nature of boundary disagreements in South and Southeast Asia and the various factors that contribute to their perpetuation. The presence of contested or poorly defined borders also plays a role in fuelling territorial disputes. If not resolved quickly, these disagreements could raise tensions and lead to physical conflict.

Conflicts in the modern world are more intricate and drawn out than ever before. There are currently approximately two billion people in conflict-affected areas, or about one third of the world's population. Global problems like environmental degradation and the trafficking of people are frequently linked to armed conflict. The Social Science Research Council states (Avis, 2020),

> "*conflicts and widespread violence have complex socio-cultural, economic, and political dimensions that operate through power networks that transcend conventional conceptual boundaries, e.g., public vs. private or local vs. national*".

It is generally agreed upon (Peace Research Institute Oslo (PRIO), 2018; that wars between states have given way to other types of aggression such as counterinsurgencies, rebellious warfare, terrorist activity, organised and massive violent crime, and unrest, all of which have increased in frequency and intensity over the past few years. As Mancini (2018), Weigand (2020), and Goodhand (2018) point out, this generally favourable trend in the "successful" decrease of disputes at national levels sometimes conceals areas of hidden, simmering, or persistent conflict within and between sub-national and cross-border regions (Mancini, 2018; Weigand, 2020; Goodhand, 2018).

Goodhand (2018, 3) notes that half the nations in Southeast Asia, as well as South Asia, are affected by subnational crises with substantial trans-border features, making these conflicts the most common and longest-lasting type of dispute in the region. Then he says

- Border and frontier regions are frequently central to the dynamics of conflict, state-building and development;
- Mainly due to state-centric analytical systems and means of working and partially in response to the inability to transform a "*borderland perspective*" into operationalizable decisions, decision-makers tend to consider borderlands as unimportant;
- Considering borderlands seriously would require considerable adjustments to development and stabilization policy/practice, which would present a huge challenge to mainstream approaches.

Asia's conflicts over land continue to threaten the region's development, security, and stability. Conflicts between nations are more common when they involve claims over land. It has been found that a variety of factors, including economic and political preferences, moral factors, and competition over limited natural resources, contribute to territorial disputes (Mancini, 2013: 1).

1.3 Typology of Conflict

When two or more sovereign states have competing claims to the same area of land or water, this is called a conflict over territory. It is not uncommon for surrounding states to sponsor or tolerate rebel movements on their territory when there is a territorial conflict (Avis, 2020). Regions and regional conflicts are the focus of multiple theoretical frameworks in foreign policy and studies of conflict and peace. The regional component provides an analytic framework for understanding issues like security and violence, as well as the mutual dependence of groups of actors (Ansorg, 2011). However, certain processes and occurrences have been poorly characterized conceptually up to this point. Even more so when discussing regional patterns of aggressive cruelty, traditional theories tend to ignore events that occur below the national or international level (Ansorg, 2011). Various factors, including the geopolitical situation (Emmers, 2009) and the involvement of non-state players (Bukh, 2020), might play a part in the varying patterns of such conflicts (Table 1.1).

The characteristics that lead to governments cooperating or escalating territorial conflicts can also be used to classify disputes (Fravel, 2008). Various factors, including the geopolitical situation (Emmers, 2009) and the involvement of non-state players (Bukh, 2020), might play a part in the varying patterns of such conflicts.

Table 1.1 Types of border conflict

Type	Description	Causes
Territorial disputes	Conflicts between states or regions over the ownership of a given area	• A country's desire to increase its power. • A political need to divert attention from existing problems. • Claimed "rights" to an area based on history, ethnicity, or geography.
Positional disputes	Conflicts over the interpretation of documents defining a boundary or the way it is shown on the ground	• Boundary was drawn at a time when accuracy was not important, E.g. river boundaries (banks, navigable channels, center, etc.) • Some groups not represented when boundary was created, E.g. colonial boundaries. • Population increases or discovery of resources make accurate boundary important.
Functional disputes	Conflicts over the national policies applied at a border	• Immigration or customs regulations, E.g. need for visas or papers to cross borders. • Land use and location policies between neighbours, E.g.Locating landfills,or polluting industries near border.
Resource disputes	Conflicts over the use of resources created or complicated by a boundary	• Resource straddles a border, E.g. River Basin, oil fields.

Source Adopted from Brunet-Jailly, 2015.

Parker asserts that borders are the dividing lines between states, and boundaries are the 'unspecific divides or separators that indicate limits of various kinds'. Borderlands are the dividing lines between states. As such, it is possible to view boundaries as having a psychological component and functioning in support of the concept of a border. As seen in the case of India and Pakistan, psychological obstacles and the consequences for borders play an essential part in the development of tensions, especially when national "pride" is tied to a conflict over the border (Tripathi & Chaturvedi, 2019). Historically asserted claims of ownership are frequently cited as the root of conflicts over land in Asia (Bukh, 2020). But instead of a long tradition per se, elements like imperial rule and its legacy, the Cold War, and the internal affairs of the various parties engaged are seen as crucial to comprehending the genesis of disputes and the ways in which they are inscribed with a new significance (Avis, 2020). Each instance of an issue of territorial integrity has its own distinct characteristics, and the development of a boundary or contention over land is highly context-dependent.

As more non-state organizations and regional and international actors are drawn into conflicts, they tend to drag on for longer. There is a growing body of evidence connecting these phenomena to larger international problems like global warming, natural calamities, cyber security, and international organized crime. British officials have identified the following as potential contributors to the development and spread of violent conflicts:

- The growing danger from terrorist attacks, fanaticism, and political unrest;
- The revival of state-based dangers and rising interstate rivalry;
- The effects of technology, particularly cyberthreats, and the broader implications of technological progress;
- The weakening of an international framework based on regulations, which makes it more difficult to work together to address global crises.

Inequality in the distribution of wealth is another major contributor. Anger and strife can arise from economic and social inequalities because of the perception of unfair treatment. The lack of access to basic resources like land and water makes these conflicts worse. In order to successfully resolve micro conflicts in the region, it is essential to address these root causes. These wars are frequently precipitated by the unfair allocation of resources and the marginalisation of some groups. Disputes on a smaller scale might be sparked by political grievances and power struggles between governments. Micro-conflicts can also be attributed to political instability and poor leadership. Marginalised communities may resort to violence in states where institutions are weak and corruption is widespread. Micro-conflicts are fuelled, in part, by economic disparity and social isolation, as those who experience these things are more likely to resort to violence as a form of redress.

1.4 Actors

There are usually many different actors, reasons, structural circumstances, and dynamics at play in conflicts between regional frameworks. Because of this intricacy, studying the dynamics of conflict in the area is a challenging task. It is not well understood how violence spreads across regions or what causes a regional conflict system to form (Ansorg, 2011). When there is a border controversy, it is not always obvious who the main players are, how they interact with one another, or how much sway they have on the dynamics and outcomes of the conflict. Border conflicts, then, involve not just the armed forces of competing nations but a wide range of global actors of violence and security engaged in intricate webs of interdependence and competition for political hegemony and monopolies of force (Avis, 2020). According to Ansorg (2011), the resulting regional dispute system can be regarded as a regionally delineated area of instability characterized by interrelated armed conflicts involving a plurality of various sub-state, national, or international actors (Ansorg, 2011).

1.4.1 Actors from the State

The activities of governments have an impact on the quality of relations between countries. Conflicts can be lessened when people from other countries work together to solve problems and reap the economic, political, and social rewards of their efforts. Similarly, the risk of war and the collapse of cooperation are made clear when countries take an adversarial stance towards one another over border disputes (Avis, 2020). It is evident that political factions and the armed forces may aggravate disputes by taking acts of aggression such as occupying contested regions (as occurs in the Cambodia-Vietnam case study) or by tolerating the illegal operations of insurgent groups (as in India's claims and the supposed operation carried out by the United Liberation Front in the region of Assam from centers in Bangladesh).

1.4.2 Actors Other than the State

There may still be state players present in these locations, but they are only one piece of the wider conflict system in which private, local, and international actors of violence take over the production of insecurity in areas where the reach of the state is weak. Such domestic aggression entrepreneurs not only prey on the populace but also charge exorbitant fees to keep the populace and their communities safe (Ansorg, 2011). Militant groups that span international borders, like the United Liberation Front of Assam, may be considered non-governmental organizations because of their alleged activities on the other side of the Indo-Bangladesh border. As a result, if one country is seen as giving radicals a free hand, it might strain diplomatic ties between the two countries.

1.4.3 Groups Based on their Ethno-Linguistic and Religious Identities

In the context of conflicts over land, issues regarding identity should not be dismissed as trivial. The act of violence can spread across borders when people within various categories of identity are related to one another, as seen in the Balkans in the 1990s. Furthermore, refugee movements and the economic and political fragility of a conflict area can produce problems in a wider region, as seen during the genocide in Rwanda and the war in Burundi in the 1990s. Borders have been crossed due to religious and cultural tensions in Southeast Asia as well as South Asia, which have escalated. Conflicts between Muslims and Buddhists in Myanmar have resulted in atrocities against Rohingya populations, while in north-east India, migrants from Bangladesh have been denied voting rights. Consequently, ethnic, linguistic, and religious communities can either exacerbate or ameliorate violent disputes (Avis, 2020).

1.4.4 Actors in the Private Security Sector

In recent years, private security businesses have mushroomed alongside terrorist groups, helping to secure valuables. Targets may emerge, for example, when a symbol of regional power, influence, or investment is attacked. In November 2018, for instance, a group of attackers in Karachi tried to break into the Chinese consulate. The assailants were all members of the Fidayeen Majeed Brigade, a newly formed group within the Balochistan Liberation Army with an emphasis on attacking Chinese targets (EFAS, 2018). When assets are at risk, such attacks can spark regional conflicts and increase the need for private protection. The goal of the terrorist attack was to draw international attention to the ongoing discrimination against the Baloch by successive authorities in Pakistan. The growing number of Chinese labourers in Balochistan, an area in Pakistan that is strategically important due to its gas and mineral resources, has exacerbated this sense of unfairness.

1.4.5 Structured Criminal/Illegal Groups

The scope of organized crime is hard to pin down and very contextual. Drug trafficking, the trafficking of people, the trafficking of weapons, and money laundering are all examples of criminal activity. Although TOC impacts all nations, the nature of illicit markets and their effects vary greatly from place to place. Even though Southeast Asia is thriving as a result of growing social contacts, trade, and investments in structures, the region faces threats from growing organized criminal networks and syndicated networks. International organised crime in Southeast Asia poses growing threats that are more intertwined with those in the rest of the area and beyond (Avis, 2020). Simultaneously, criminal organizations based in Southeast Asia have expanded their operations on a global scale, facilitating the trade of enormous quantities of highly lucrative methamphetamine, shipments of wildlife and natural resources, and an ever-expanding selection of fake consumer and industrial goods (UNODC, 2019). They also persist in human trafficking for oppression in both the sexual and labour markets, as well as migrant smuggling.

In the Southeast Asian region, criminal organization gangs have amassed unprecedented power and are in charge of "industries" with annual revenues in the billions. The rule of law is undermined and corroded by the financial strength of multinational organized criminal networks. Southeast Asia's diverse population, language and culture, and island and hilly topography all combine to make government in the region particularly difficult (Avis, 2020). In addition, the region is made up of developed, developing, and least developed countries, which not only slows down the development of a unified regional market but also has a significant impact on the circulation patterns of criminal organizations. Crime syndicates have centralized their activities in nations that provide a more favourable business climate, such as those that lack the resources to carry out and regulate effective countermeasures against international criminal activity (UNODC, 2019).

1.5 The Impact of Micro-Conflicts on the Region

Micro-conflicts affect not just the countries directly involved but also neighbouring states, often with far-reaching implications. Populations are displaced because of these wars, leading to a humanitarian catastrophe and putting a burden on the resources of the towns that take them in. Governments are having a hard time enforcing law and order due to the presence of armed groups and the proliferation of small weaponry. The victims of violence and displacement are almost often innocent civilians, especially those who live in locations where conflicts are ongoing. In addition, micro-conflicts have dire economic consequences, which stymie growth and deepen poverty. The micro-conflicts can have a spillover effect, destabilizing neighbouring countries and stoking regional tensions. If South and Southeast Asia are to enjoy lasting peace and prosperity, it is crucial that the effects of these conflicts be acknowledged and that steps be taken to end them. They obstruct commerce, scare off foreign investment, and take money away from things like education and healthcare.

Micro-conflicts are also disruptive to regional integration and economic growth. As a result, they discourage business and investment and foster an environment of instability. The resulting economic stagnation adds gasoline to the flames of discontent, heightening the prospect of more war. In addition, regions afflicted by micro-conflicts can be fertile ground for extremist ideas and terrorist activities. Furthermore, the security of surrounding areas might be negatively affected by micro-conflicts. They cause instability in the surrounding countries by providing a breeding environment for extreme beliefs and non-state armed groups. Spillover violence and population relocation can also put a burden on already limited resources and exacerbate tensions between host and refugee communities. Therefore, settling these smaller conflicts is not only a moral need but also crucial for maintaining peace and order in the region.

1.6 Case Studies of Micro-Conflicts in South and Southeast Asia

Several studies focusing on conflicts in South and Southeast Asia shed light on the complicated and varied nature of these conflicts. These examples show the difficulty and complexity of resolving small conflicts in South and Southeast Asia. Myanmar's

Rakhine state, where the Rohingya minority has been persecuted and subjected to violence, is one such example of a continuous conflict. Many Rohingya have fled to neighbouring Bangladesh as a result of this fighting, creating a significant refugee crisis there. Religious and ethnic tensions, economic and social injustices, and political challenges all play a role in fuelling this complicated war. Resolving this conflict entails addressing the underlying causes of prejudice and marginalisation, as well as promoting interfaith dialogue and inclusive governance. Similarly protracted is the war between India and Pakistan over the contested territory of Kashmir. Multiple conflicts and persistent tensions between the two nuclear-armed nations have resulted from this geographical issue. A lack of trust and competing political agendas have stymied diplomatic efforts to find a peaceful resolution to the crisis.

Another case in point is the insurgency in India's North-Eastern states, where a number of ethnic and separatist parties have been fighting for independence for some time. This ongoing war has slowed progress in the area and put a strain on India's armed forces. The war between the Tamil minority and the government of Sri Lanka has also resulted in widespread human suffering, including death. Another illustration is how ethnic and religious tensions have fuelled the insurgency in Thailand's southern provinces, making peace difficult to achieve. These examples illustrate the necessity for systems-level strategies to deal with systemic issues and localised disputes. In Southeast Asia as well as South Asia, disagreements regarding borders continue to be a contentious political issue, shaping the nature of interactions between countries. Traditionally, the borders in this area have been "soft" and "porous" (Goodhand, 2018). In order to properly analyse the state of regional security and conflict in South and Southeast Asia, a more in-depth knowledge of the dynamics of border disputes and conflicts is required. Uncertainty surrounds geopolitical changes, the depletion of natural resources, and ecological degradation in Asia (Mancini, 2013). The South China Seas are a potential hotspot due to the convergence of multiple challenges that threaten peace and stability in the region and beyond (Mancini, 2013). There is reason to be hopeful that territorial conflicts can be settled peacefully due to the region's ongoing trend towards cooperation, the increasing importance of regional organisations (ASEAN, etc.), and the developments of democratisation (Avis, 2020).

Colonialism's lingering effects can be seen in the region's several boundary disputes, as the designation of territorial boundaries by colonial rulers frequently ignored the ethnic origin and cultural traditions of the local inhabitants in favour of simplicity, power sharing negotiations, and pacts (Mishra, 2016). The following list of conflicts is meant to illustrate the various causes of war rather than serve as an exhaustive catalogue of all possible conflicts. These conflicts were chosen to illustrate how local tensions can quickly escalate into international crises when they interact with global struggles for political or economic control, are stoked by political players eager to capitalize on discontent, or are compounded by limited resources. Several of the conflicts we'll be looking at below may be traced back to decolonization and other regional and internal disagreements that were incorporated into the Cold War (Avis, 2020). These disagreements remain significant in the ties between them and other countries, notwithstanding the end of the Cold War and the subsequent radical regional and global developments. Indeed, as time goes on and contexts shift, new meanings are being imprinted on old disputes.

1.6.1 Conflicts between India and Bangladesh

India and Bangladesh share over 4096.7 km of border, making it the fifth-longest land boundary in the world. New Delhi is worried about insurgent organizations, smugglers, and illegal immigrants using the fragile border between India and Bangladesh, so it has been working to increase security along this border (Avis, 2020). Making remarks on the North Eastern border areas more generally, Lakshman and Jha (n.d.) point out a few trends. These include the development and globalization of well-organised criminal networks with strong political influence and support and an expanding system of organizations for the communal movements of the migrants, especially through an expanding complex of madrassas. Poorly delineated border regions, intermarriage across boundaries of ethnicity, and the presence of non-native populations all contribute to escalating tensions along the line of control (Avis, 2020). The broadness of the boundary, the ruggedness of the landscape, and the severity of the weather all make for challenging surveillance. Hazarika believes that the consistency of migrant mobility lends it an essential, even deadly, edge, particularly given that such movements take place in a region already disturbed by conflicts. While the report focuses on conflicts that occur between countries, violence can also be seen within countries, but between different sub-national groups. When people's ethnic, linguistic, or religious ties span international boundaries, it can cause tension on a regional level (Baruah, 2005).

1.6.2 The Conflict between Nepal and India

The free flow of individuals and products beyond state boundaries in the South Asian region is severely regulated. But there are a few notable outliers. The international border between India and Nepal is open and unimpeded for the most part, allowing for free trade and travel between the two countries. India's current Prime Minister, Narendra Modi, paid a historic visit to Nepal in 2014. This greatly boosted ties between the two countries. During Modi's visit, several significant agreements were signed, revitalizing ties between India and Nepal. For the purpose of carrying out development and infrastructure initiatives, India extended a US$1 billion soft line of credit (Tripathi, 2019). Border concerns have occasionally become contentious despite generally amicable relations. The Madheshi movement was a recent example of tensions between India and Nepal connected to their shared border. A protesting subset of the Madheshi ethnic group in Nepal shut down trade between their country and India in 2015. Because of the Madheshi people's cultural and historical connection to India, the closure of the Indo-Nepali borderline has strained relations between the two countries (Tripathi, 2019).

Madhesh, in southern Nepal, is a lowland area that borders India on two sides. A large percentage of the Madheshi people call the Indian-Nepali border region home. There is a strong bond between this group and those in northern India. Disgruntled community leaders voiced their disapproval of the new Nepalese constitution, which stoked tensions in the area. India has sounded the alarm at this time out of fear that the unrest among the Madheshi population would spread to Indian Territory, given the closeness between the two countries. In an effort to warn Nepal of impending unrest within the Madheshi people, India made diplomatic overtures in 2019 (Tripathi, 2019). When the Nepali government ignored the protesters' demands, the situation got worse. Protesters in Madhya Pradesh shut down the border between India and Nepal in September 2015, escalating the unrest. Because Nepal remains landlocked and relies so much on India for international commerce transit, the blockade has had a significant impact on Nepalese society at large (Avis, 2020). The government of Nepal has blamed India for stoking the turmoil by providing tacit support for the Madheshi community's blockade. The Nepalese government has called the current impasse an "unofficial blockade by India. Cracks appeared in the Nepalese government as a result of the blockade. Nepalese perceptions of India were also affected.

The relationship between China and India has matured to the point that it is now a geopolitical issue. To be sure, any Chinese investment in Nepal will be perceived as a move against India. Negative talk towards India and Nepal is more common than favourable. After the blockage was lifted, life went back to normal. Although the border between India and Nepal has not been altered formally, attitudes about it have. According to Tripathi (2019), an invisible line has been drawn between the two nations' minds (Avis, 2020). The lack of security personnel along the border between India and Nepal is also no longer an issue. The security conundrum that has come to define South Asian border politics more broadly now applies to the frontier between India and Nepal.

1.6.3 Jammu and Kashmir as a Source of Controversy between India and Pakistan

There are other territorial conflicts between India and Pakistan, but the one over Kashmir is the most contentious. Kashmir is divided into thirds, with Pakistan governing one third and India the other two thirds. It is the greatest territorial dispute in the world, both in terms of the area of land at issue (over 300,000 square kilometers) and the number of people (about 17.5 million) living there (Bukh, 2020). In addition to having a significant impact on regional connections and the strategies of global powers like the United States, China, and Russia, the controversy over Kashmir has been directly responsible for thousands of fatalities and is a major source of confrontation between two nuclear. Both the lasting effects of British colonial authority in India and the circumstances of the British exit from the Indian subcontinent in 1947–1948 are at the root of the problem (Avis, 2020). There are often border clashes that culminate in civilian and military casualties. The two governments met from 2004 to 2007 to figure out a plan to settle the conflict, but the talks were halted after the terrorist attacks in Mumbai in 2008 (Avis, 2020). The territory of Kashmir is contested between two governments. Pakistan, however, eventually acknowledged a struggle for autonomy in Kashmir's highlands and the impossibility of incorporating all of Kashmir within its boundaries. Similarly, it would appear that India has no intention of annexing the parts of Kashmir that are currently under Pakistani authority. The likelihood of a resolution to the issue in the near future remains low despite these significant shifts in the stances of the two states (Blarel and Ebert, 2015; Bukh, 2020).

The construction of the Baglihar project over the River Chenab in Kashmir, which is administered by India, is a prime illustration of how rivalry for commodities may increase conflicts along the Indo-Pak borderline (Avis, 2020). The Indus Valley Waters Treaty, signed between India and Pakistan in 1960, was used to avoid a water war between the two states by granting India exclusive rights to the waters of the Eastern Rivers up until the point when they flow into Pakistan and granting Pakistan the rights to the waterways of the Western Waterways. Pakistan has always been at India's mercy when it comes to the Indus River because India regulates its level upstream (Bukh, 2020). Changing patterns of rainfall due to the effects of climate change have led the United Nations to label Pakistan as a "water scarce" country (Werleman, 2020). India's increased authority over Kashmir, along with decreased water flow from hydroelectric power initiatives, could lead to a major confrontation between India and Pakistan in the near future. By 2050, it is predicted that 40 percent less water will be available for consumption by people, raising the prospect of water-related conflicts.

1.6.4 Challenging Issues between India and China

The border between China and India is the greatest unofficial boundary in the whole world. Their disagreement concerns the borderlands of Aksai Chin and Arunachal Pradesh. Over 38,000 square kilometers in area, Aksai Chin is largely deserted. Both China, as part of its Xinjiang Autonomous Area, and India, as a portion of Jammu and Kashmir, lay claim to this area. Located in India's far northeast, Arunachal Pradesh is also known as South Tibet in China. It is home to around 1.2 million people. China claims 90,000 square kilometers (about 36,000 square miles) of land in Arunachal Pradesh's eastern borderline region (Brunet-Jailly ed., 2015). There is a connection between the Kashmir conflict between India and Pakistan and the conflict over Aksai Chin. Kashmir did not have a well-defined boundary when British authority in the region came to an end. Indian official documentation from the years after India's independence showed an 'undefined' border between India and China in this area.

When China invaded and captured Tibet in 1951, it gained control of this area. India's claim to Aksai Chin is based on the boundaries established by the Dogra kings of Kashmir before colonization (Brunet-Jailly ed., 2015), which they asserted only in 1953.

Countries began increasing their military presence in the Arunachal Pradesh region during the years 1986–1987, heightening tensions in the region. In today's world, ties between India and China are generally positive and play an important role in maintaining peace in the region. The 1987 incident ushered in a period of icy calm between the two Asian superpowers, during which time both sides formally maintained their territorial claims while respecting the status quo. The Indo-China region is still prone to sporadic outbreaks of violence. The Sino-Indian boundary has been the site of violent abuses, face-offs, and conflicts between Chinese and Indian troops since May 5, 2020 (Avis, 2020). This includes areas surrounding the challenging Pangong Lake, which is located in Ladakh and the Tibet Autonomous Region, as well as the boundary between Sikkim and the autonomous region known as Tibet. The actual control line (LAC) in eastern Ladakh was also the site of more skirmishes (Avis, 2020).

1.6.5 Conflict(S) between Bangladesh and Myanmar

The border conflict between Myanmar and Bangladesh is an example of the geopolitical factors at play. Located between China and India, with access to the Bay of Bengal and the Indian Ocean, Myanmar is in a prime spot. Only one land route interconnecting East Asia, Southeast Asia, and South Asia passes through this single point. Because of its strategic position, it has attracted attention from key international actors like China, Japan, the USA, the Indian subcontinent, and the European Union. Trade between China and India is facilitated by Myanmar's ports on the Bay of Bengal and the Indian Ocean. Oil, gas, power, regional linkages, and unique economic zones are only some of the development programmes in which they have invested heavily in Rakhine state. Bangladesh and Myanmar have less than friendly ties because Bangladesh has not reaped many benefits from Myanmar's natural riches, geopolitical setting, industry, or investing activities (Avis, 2020).Bangladesh's lengthy boundary with Myanmar makes it a very close neighbour. It's rich in resources including oil and gas as well as timber, fisheries, arable land, and other minerals.

Bangladesh and Myanmar have been in an ocean border controversy for quite some time. In the Bay of Bengal, Bangladesh and Myanmar are at odds over who gets what in terms of maritime waters, the continental shelf, and special economic areas. The ocean border between Bangladesh and Myanmar was decided upon by the International Tribunal for the Law of the Sea (ITLOS) on March 14, 2012. This ruling was a major victory for Bangladesh in the courtroom. The tribunal's ruling will allow Bangladesh to begin oil and gas exploration in the Bay of Bengal, where massive reserves are predicted to exist (Ahamed et al., 2020). Myanmar has become a major centre for the manufacture and sale of methamphetamine and its derivatives, adding to the region's growing criminal activity. According to Ahamed et al. (2020), a sizable population is participating in this illegal trade along international borders. The methamphetamine and caffeine hybrid known as "yaaba" has become a serious problem in some parts of Bangladesh, most notably the city of Cox's Bazar. This inexpensive drug trade has even reached the cities of Chittagong and Dhaka, both in Bangladesh (Avis, 2020).

Law enforcement authorities have been accused because of the lucrative nature of the yaaba trade in Bangladesh. Some have even gone so far as to say that they are the main criminals and benefactors of this trade. Many innocent persons are said to have been victimised by police officers through kidnapping, cross-firing, persecution, and the filing of fake cases; yaaba is then used as a means to extract ransom (Alam, 2017). The Arakan Rohingya Salvation Army (ARSA) is said to be working out of and recruiting Rohingyas from refugee camps (Avis, 2020) adding a cross-border aspect to the national ethnic war within the Republic of Myanmar. Militant organisations in ARSA and Myanmar can connect to make the war between them more widespread. In a purportedly published statement, Al-Qaeda on the South Asian continent urged the Muslims of Bangladesh to aid the Rohingya. Whether or not Al-Qaeda is connected to ARSA is unclear. However, there are extremist linkages in Bangladesh and Myanmar that are allowing the ARSA and hardline Buddhists to expand their influential power in the region (Avis, 2020).

1.6.6 Conflicts in the South China Sea Regarding Maritime Territories

Border and maritime conflicts in the South China Sea are among the most complex in the area, if not the world (Bukh, 2020). Petroleum and natural gas are abundant in the contested regions, and half of global traffic passes through these oceans, making them strategic. These disputes influence Japan, the US, and the claimants' foreign policies. The disputed waters overlap with those of China, Taiwan, Brunei, the Philippines, Vietnam, Indonesia, and Malaysia. Access to fisheries is crucial in the conflict between Hainan and Guangdong in China, Vietnam, and the Philippines (Avis, 2020). The South China Sea is a world-class fishing spot with rich biodiversity. Sea fishing accounts for 10% of the world's fishery harvest, making it vital to fishing nations' economies. China is reluctant to compromise over its territorial claims due to its rising need for energy resources, notably oil and gas, to sustain its development and its ambition to lessen its dependence on Middle Eastern oil (Bukh, 2020). Due to tensions, most hydrocarbon estimates in the Spratly Islands are unverified. International shipping is geostrategic in the South China Sea. The South China Sea transports most energy and raw materials from the Malacca Straits to China and Japan. The Chinese People's Liberation Army Navy (PLAN) has strengthened its sea capabilities due to the sea's strategic importance. The U.S. values open and stable maritime commons for international trade and prosperity; hence, it protects the area's maritime routes (Avis, 2020).

Due to the ambiguity surrounding China's legal rights and potential attempts to demonstrate sovereignty in regions that are too far from its ports to be parts of its EEZ, tensions have arisen between China and other claimants. Violence in the area is exacerbated by Chinese domestic politics as well. Problems in the South China Sea have been worsened by the fact that several Chinese government entities have conflicting goals and rarely work together. The problem is made worse by the fact that the Chinese navy has used maritime conflicts to justify its upgrades and by the fact that nationalist sentiment around territorial disputes only serves to exacerbate the situation (Avis, 2020). The three countries with the strongest and most public claims to the South China Sea are China, Vietnam, and the Philippines. If the former decides to claim the entirety of the EEZs surrounding the Spratly Islands, there will be significant overlap with the EEZs that the Philippines, Brunei, Malaysia, and Vietnam have claimed (Avis, 2020), which could have significant consequences for all of the claimants. Beijing may endeavour to achieve this objective, but it is also considering asserting "historical rights" to other marine areas within the nine-dash line (Bukh, 2020).

There is a high potential for armed conflict due to the increasing number of law enforcement and paramilitary vessels operating in uncharted waters near contested territories. Recently, they have been at the centre of tensions between China and the Philippines, as in the Scarborough Reef standoff in April 2012. Warfare in the South China Sea has its roots in international politics. Clinton reiterated in 2010 that free passage in the South China Sea was a U.S. national interest, and Trump has been vocal in his opposition to China. According to Chinese observers, the United States is used by ASEAN nations as a buffer against China's growing might, while the United States has been using ASEAN nations to increase its presence in the region (Bukh, 2020). Bukh (2020) claims that the Chinese government employs a well-coordinated strategy to expand its regional influence and contain other powers. By appealing to economically weaker ASEAN members like Cambodia, China is able to maintain regional superiority and further exacerbate divisions over the South China Sea dispute. The Cambodian capital of Phnom Penh is being forced to accept China's regional activities despite the destruction of ASEAN due to territorial disputes with its fellow ASEAN member, Cambodia. As Kizekova and Gerstl (2017) point out, ASEAN involves China in regional institutionalism and practices multitrack diplomacy in order to defuse any lingering concerns (Avis, 2020).

1.6.7 Conflicts Between Thailand and Cambodia

There are several border disputes between Thailand and Cambodia, but the most contentious one is over an area of land measuring about 4.6 square kilometers that encompasses the Hindu temple of Preah Vihear (known in Thai as Phra Viharn). The conflict has sparked a number of military confrontations. The Preah Vihear region in Cambodia is home to the sanctuary, which is perched on a cliff in the Dangrek/Donrak mountain range and has been claimed by Thailand as a member of its Sisaket Region. According to the Institute for Peace and Conflict Studies (2012), the controversy is fueled by concerns over national identity and sovereignty (Avis, 2020). After hostilities between Thailand and Cambodia escalated in 1954, when the Thai military took the temple, the matter was taken to the International Court of Justice (ICJ) in 1959. Cambodia now legally owns the temple because of a 1962 International Court of Justice judgement. Despite Thailand's concerns, the World Heritage Committee inscribed the temple as a UNESCO World Heritage Site in 2008 (Wagener, 2011). This renewed the controversy. Cambodia petitioned the ICJ in April 2011 to have its 1962 judgement reinterpreted and to issue an interim restraining order.

Cambodia claimed that the 1962 ruling gave Cambodia control of the temple and its surrounding area. In response to Cambodia's allegations, Thailand contended in a submission to the International Court of Justice (ICJ) that the pervasive problem was not over possession of the temple according to the 1962 judgement but rather a new border controversy over the 4.6 sq. km related to the temple (Bukh, 2020). The dispute was contextualised in the larger discussion of Thai territory lost to British and French colonial rulers in the late 19th and early 20th centuries by the Thai irredentist paradigm that gained traction in 2008. This argument has been attributed to colonialism and, more especially, the strategies left behind by the colonial rulers in Southeast Asia (Avis, 2020). Although the current conflict between Thailand and Cambodia can be traced back to the region's colonial era, internal conflicts in Thailand were the primary cause of the intensification of the problem in 2008 and the ensuing confrontations (Avis, 2020). In the wake of the internal conflict between the "royalists" and the followers of ex-Prime Minister Taksin Shinawatra, the problem was rekindled and politicised in Thailand at that moment (Bukh, 2020). Thus, it is clear that territorial conflicts can be used as a weapon in regional politics, either as an expression of authority to a regional audience or as proof of nationalist claims.

1.6.8 Borderland Conflict Between Vietnam and Cambodia

The recent outbreak of violence near the border between Vietnam and Cambodia is a reminder that unrest can arise even in seemingly peaceful regions. In 2006, the prime ministers of the two countries reached an agreement to settle the boundary dispute between Moc Bai and Ba Vet. It was supposed to be finished in 2012, but there are still unidentified sections along the 1200-kilometre boundary (Cotillon, 2017). As anti-Vietnamese nationalist sentiment and imagined Vietnamese threats have gained traction in recent years, conflicts along the Vietnamese-Cambodian borderline have also increased. According to reports, conflicts between Vietnamese and Cambodian residents erupted into deadly fights in 2009 after Sam Rainsy, head of the opposition Cambodia National Rescue Party (CNRP), allegedly pushed locals to dismantle border markers in the Svay Rieng region (Avis, 2020). Violent fights broke out between Vietnamese and Cambodian border guards in June 2015.

Legislators from the CNRP, Real Camerin and Sam, instigated the violence by leading a group of about 250 Cambodian demonstrators armed with sticks into Vietnamese land. Both Vietnamese security personnel and locals actively opposed their intrusion. Dozens of Vietnamese and Cambodian nationals were hurt when the argument turned violent (Cotillion, 2017). Another time, entrenched regional interests are using the junction of national political narratives to their advantage, fueling instability in the area at large. Given the delicate nature of border areas, wars there can have a chilling effect on national unity and sow the seeds of dispute.

1.6.9 Conflict over Territorial Boundaries Between China and Vietnam

These countries include the People's Republic of China (PRC), the Republic of China (ROC), Brunei, Malaysia, the Philippines, and Vietnam. The Spratly Islands, the Paracel Islands, Scarborough Shoal, and many other areas in the Gulf of Tonkin are just some of the islands, reefs, banks, and other features in the South China Sea that are at the centre of the conflicts. It is also disputed that the waters close to the Indonesian Natuna Islands are not part of the South China Sea. Disputes over maritime territory aren't limited to the South China Sea; the East China Sea is likewise fraught with tensions over the contested Senkaku Islands and Socotra Rock (Keck, 2014). The land border between China and Vietnam is 1297 kilometres (806 miles) long and runs from a tripoint with Laos in the west to the Gulf of Tonkin coast in the east; the maritime border runs via the Gulf of Tonkin and the South China Sea. Although a land boundary treaty has been signed to settle disagreements on the land side of the border, disagreements over the Spratly and Paracel Islands' sovereignty have left the marine border unsettled.

Since 1974, when Vietnam first claimed China of seizing the Paracel Islands for economic gain, the two countries have been at odds over maritime territory in the South China Sea (Avis, 2020). The Paracel and Spratly islands are at the centre of the territorial conflict between China and Vietnam in the South China Sea. In 2009, in response to a UN Note Verbale criticizing China's claims, Vietnam and Malaysia issued their own Note Verbale. According to China's Declaration Verbale submitted to the United Nations in 2009, Beijing indisputably controls the islands in the South China Sea and the area of water immediately surrounding them as well. In the South China Sea, several countries have territorial conflicts owing to competing island and marine claims (Keck, 2014).

1.7 The Role of International/Regional Organizations in Resolving Micro-Conflicts

Micro-conflicts in South and Southeast Asia are typically resolved through international organisations. They serve as a meeting place for conflict resolution specialists and mediators. Such as the United Nations' participation in peacekeeping missions and in mediating agreements between warring parties. Its presence and objectivity can pave the way for trust to be established and conflict to be resolved. International organizations have played a crucial role in facilitating the resolution of boundary conflicts in South and Southeast Asia. Organizations such as the United Nations (UN), the Association of Southeast Asian Nations (ASEAN), and the International Court of Justice (ICJ) have provided platforms for dialogue, mediation, and arbitration (International court of...). Efforts to resolve conflicts have been supported by a wide range of institutions, including the United Nations, the Association of Southeast Asian Nations (ASEAN), and the South Asian Association for Regional Cooperation (SAARC). As a meeting ground for disputants, these groups facilitate communication, mediation, and compromise. Their participation and backing lend legitimacy to the peace process and encourage the warring parties to negotiate in good faith. ASEAN, as a regional organization, has also played a significant role in promoting stability and cooperation in the region. Through its mechanisms such as the ASEAN Regional Forum (ARF) and the ASEAN Regional Centre for Arbitration (ARCA), ASEAN has fostered dialogue and facilitated discussions on boundary disputes among its member states (Castro, 2017). These groups can also supply the necessary resources and technical support to ensure the successful implementation of peace building initiatives. If permanent peace is to be established in the region, cooperation between local and international actors must be bolstered.

Communities in need also receive help and humanitarian aid from foreign organisations. They aid in the reconstruction of war-torn areas by relocating and rehabilitating displaced people. In addition, institutions like the Asian Development Bank (ADB) give financial aid and develop programmes that aid in peace-making. In addition, they aid in promoting good governance and strengthening institutions to forestall the outbreak of future conflicts by facilitating capacity-building activities. The UN, through its various agencies and specialized bodies, has been instrumental in promoting peaceful resolution and facilitating dialogue between conflicting parties(United Nations, 2018). The UN Charter's principles of peaceful settlement of disputes and respect for territorial integrity have provided a framework for addressing boundary disagreements and micro-conflicts. The ICJ, as the principal judicial organ of the UN, has been involved in resolving boundary disputes through legal means(United Nations, 2018). Its impartiality and expertise in international law have provided a forum for countries to seek legal recourse and settle disputes in a peaceful manner.

1.8 Strategies for Resolving Micro-Conflicts

An all-inclusive, multi-faceted strategy that targets the sources of micro conflicts and encourages long-term stability is necessary to end these disputes. Prioritising conversation and diplomatic efforts as means of resolving conflicts is a crucial tactic. Trust, understanding, and agreeable solutions can all be achieved through open communication among all parties involved. De-escalating emotions and coming to an amicable agreement can also greatly benefit from mediation and facilitation by impartial third parties. Economic growth promotion in post-conflict zones can help restore stability. Addressing socioeconomic disparities and easing tensions can be accomplished through expanding access to educational and healthcare resources and expanding work possibilities. South and Southeast Asian countries can help pave the path for peaceful resolutions to micro-conflicts by prioritising discussion and diplomacy.Promoting economic development as a means of achieving peace is another crucial tactic. Reducing the potential for violence requires addressing economic inequality and providing opportunity for marginalized people to help ease grievances. Long-term peace and security can benefit from investments in infrastructure, schools, and new jobs. Targeted development programmes that aim to reduce poverty and inequality, two major contributors to micro-conflicts, can serve to assuage grievances and open doors for underprivileged groups. Integration and collaboration on the economic front between warring parties can also help bring about lasting peace.

Legal frameworks and arbitration mechanisms are essential tools for resolving boundary disputes in South and Southeast Asia. These mechanisms provide a structured and impartial approach to resolving conflicts, ensuring that disputes are settled based on established legal principles and international norms. One such mechanism is the use of international law, including the United Nations Convention on the Law of the Sea (UNCLOS), which provides a legal framework for resolving maritime boundary disputes. UNCLOS establishes the rights and responsibilities of states in their use of the world's oceans and provides guidelines for delimiting maritime boundaries. Arbitration has also been utilized to resolve boundary disputes, with conflicting parties voluntarily submitting their disputes to an impartial tribunal. The involvement of third-party arbitrators helps mitigate power imbalances and ensures a fair and unbiased resolution.

Supporting economic growth can be a potent means of fostering stability in South and Southeast Asia. The underlying causes of wars and other violent conflicts can be addressed by investing in infrastructure, job creation, and poverty reduction. Especially for underprivileged groups, economic opportunity might offer an alternative to extremism and violence. There is less need for war when countries are economically dependent on one another. The area can set the stage for long-term stability if it prioritizes equitable and sustainable development.Micro-conflicts can only be resolved through a combination of international mediation and regional collaboration. Trust, cooperation, and mutual understanding among member states can be bolstered by strengthening regional structures like ASEAN and SAARC. These groups can pave the way for amicable conflict resolution by facilitating confidence-building measures and fostering cross-border collaboration. Trust and cooperation between countries in close proximity can be bolstered by setting up systems for sharing information, collaborating on security measures, and working together on development projects. Regional cooperation can help bring about enduring peace in South and Southeast Asia by addressing similar concerns and supporting shared interests.

1.9 Challenges and Obstacles to Resolve Boundary Disagreements

Resolving boundary disagreements in South and Southeast Asia is not without its challenges and obstacles. The complex nature of these disputes, coupled with historical animosities and competing interests, often hampers progress towards a peaceful resolution. One significant challenge is the deeply entrenched historical and emotional attachments to disputed territories. Nationalistic sentiments and public opinion often complicate negotiations, making it difficult for governments to reach mutually acceptable solutions.

Another significant challenge is the lack of political will and commitment from conflicting parties. In some cases, domestic political considerations and nationalist sentiments can overshadow the importance of resolving boundary disputes. This can make it difficult for leaders to make concessions or engage in meaningful dialogue, perpetuating the status quo.

One more obstacle is the presence of non-state actors and external influences. In some boundary disputes, non-state actors such as militant groups or separatist movements complicate the resolution process by engaging in acts of violence or undermining diplomatic efforts. External powers, seeking to advance their own interests, can also exacerbate conflicts by providing support to one side or engaging in provocative actions.

Furthermore, geopolitical rivalries and power dynamics among major regional players can hinder progress in resolving boundary conflicts. The influence of external actors, such as superpowers, can exacerbate tensions and undermine diplomatic efforts. Balancing competing interests and ensuring the participation of all relevant stakeholders is crucial for successful conflict resolution.

1.10 Future Prospects for Resolving Boundary Disagreements and Micro-Conflicts in the Region

Despite the challenges, there are reasons to be optimistic about the future prospects for resolving boundary disagreements and micro-conflicts in South and Southeast Asia. The increasing interconnectedness of the global community, the rise of regional organizations, and the growing recognition of the importance of peaceful resolution all contribute to a more conducive environment for dialogue and negotiation. Platforms like ASEAN provide a forum for addressing these issues and have shown the potential for effective conflict resolution.

It is crucial for countries in the region to prioritize regional cooperation and work towards building trust and understanding. By engaging in sustained dialogue, fostering people-to-people contacts, and promoting economic integration, countries can create an environment that is conducive to resolving boundary disputes. Moreover, advancements in technology and scientific methods have improved the accuracy of boundary demarcation and resource assessment. The availability of precise data and mapping techniques enables nations to make informed decisions, reducing the potential for misunderstandings and disputes.

Furthermore, the role of international organizations and legal frameworks should be strengthened to provide a more robust mechanism for dispute resolution. By enhancing the mandate and capacity of organizations such as the UN and ASEAN, and ensuring adherence to established legal principles, the region can make significant strides towards peaceful resolution. Building on successful case studies and leveraging international legal frameworks, arbitration mechanisms, and the expertise of international organizations, South and Southeast Asian nations can work towards peaceful resolutions of boundary disagreements. By prioritizing dialogue, cooperation, and the shared goal of regional stability, these nations can foster an environment conducive to lasting peace and prosperity.

1.11 Conclusion

Therefore, it is clear that resolving micro-conflicts in South and Southeast Asia is a challenging task. Peace, however, can be maintained indefinitely if the reasons of these disputes are identified, their effects on the region are evaluated, and the right solutions are put into place. Cooperation among governments, international organizations, and civil society is essential for facilitating communication, resolving conflicts, and advancing economic growth. The area can prepare the road for future peace by learning from the complexities of previous conflicts, pursuing dialogue and diplomacy, encouraging economic development, and strengthening regional cooperation. Cooperation and dedication to long-term solutions from both domestic and international parties is essential. South and Southeast Asia can only attain permanent peace and development via concerted, regional efforts. Peacebuilding in South and Southeast Asia requires preemptive measures, negotiated settlements, and post-war reconstruction. All stakeholders, especially traditionally marginalized groups, should be represented in this road map's decision-making procedures, which should be emphasized. Furthermore, it should stress the significance of fostering trust, encouraging conversation, and resolving problems' root causes. South and Southeast Asia's people can have a peaceful and successful future if these suggestions are implemented and an all-encompassing strategy is pursued. In order to maintain peace and harmony in the region, all parties involved must work together to fund peace-building initiatives and find solutions to minor issues. A path including continued discussion, inclusive government, economic empowerment, and regional collaboration is necessary to achieve enduring peace. South and Southeast Asia can overcome the hurdles and pave the way for a peaceful and prosperous future by addressing the underlying causes of micro-conflicts and establishing an environment of trust and collaboration.

In conclusion, resolving boundary disagreements and micro-conflicts in South and Southeast Asia is crucial for regional stability, economic development, and cooperation. These disputes have deep historical roots and arise from a variety of complex factors, making their resolution a challenging task. But, through sustained diplomatic efforts, engagement with international organizations, and adherence to established legal frameworks, the region can overcome these challenges and work towards a future of peace and cooperation. It is essential for countries in the region to prioritize dialogue, compromise, and the peaceful settlement of disputes. Through diplomatic initiatives, negotiations, the involvement of international organizations, legal frameworks, and arbitration mechanisms, there have been successful cases of resolving these conflicts. However, challenges and obstacles persist, requiring sustained efforts and political will. By doing so, they can pave the way for increased regional integration, economic prosperity, and a more stable and secure future for South and Southeast Asia as a whole. By prioritizing peaceful resolution and regional cooperation, South and Southeast Asia can pave the way for a more stable, prosperous, and integrated future.

References

Ahamed, A et al. (2020). Bangladesh-Myanmar border relations: A study of some Geopoliticaland economic issues. *European Scientific Journal 16*, (22). https://doi.org/10.19044/esj.2020.v16n22p320, [Accessed on-29/09/2023].

Alam, A. (2017). India, nonalignment and emerging global governance. *India Quarterly: A Journal of International Affairs, 73*(3), 273–291. https://doi.org/10.1177/0974928417716210. (Original work published 2017).

Ansorg, N. (2011). How does militant violence diffuse in regions? Regional conflict systems inInternational relations and peace and conflict studies. *International Journal of Conflict and Violence, 5*(1), 173–187. https://core.ac.uk/download/pdf/276528767.pdf, [Accessed on- 29 Sep 2023]

Avis, W. (2020). Border disputes and micro-conflicts in south and Southeast Asia, Helpdesk Report, University of Birmingham.

Baruah, S. (2005). *Durable disorder: Understanding the politics of northeast India* (p. 265). Oxford University Press.

Blarel, N., & Ebert, H. (2015). Explaining the evolution of contestation in South Asia. *International Politics, 52*. https://link.springer.com/article/10.1057/ip.2014.44

Brunet-Jailly, E. (2015). Border disputes: a global encyclopedia. BC-Clio.

Bukh, Alexander. (2020). Territorial disputes in Asia: Colonialism, cold war and domestic politics. in: Inoguchi, T. ed. The SAGE handbook of Asian foreign policy. *Sage,* published on-January, 2020. https://www.researchgate.net/publication/338633837_Territorial_Disputes_in_Asia_Colonialism_Cold_War_and_Domestic_Politics, [Accessed on-20 Aug 2023].

Castro, Renato Cruz De. (2017). The ASEAN Regional Forum in the Face of Great-Power Competition in the South China Sea: The Limits of ASEAN's Approach in Addressing 21st-Century Maritime Security Issues? https://www.kas.de/c/document_library/get_file?uuid=43dafd95-a5cc-e07e-6d14-73ac2d12cda6&groupId=288143, [Accessed on-29 Sep 2023].
Cotillon, H. (2017). Territorial disputes and nationalism: A comparative case study of Chinaand Vietnam. *Journal of Current Southeast Asian Affairs, 1*(36), 51–88. ISSN 1868-1034. http://eprints.lse.ac.uk/81185/, [Accessed on- 29 Sep 2023].
EFAS (2018). Attack on Chinese consulate in Karachi: Wake-up call for Pakistan and China toaddress grievances of the Baloch. EFAS. https://www.efsas.org/commentaries/attack-on-chinese-consulate-in-karachi-wake-up-call-for-pak-and-china-to-address-grievances-of/, [Accessed on-29 Sep 2023].
Emmers, R. (2009). Comprehensive security and resilience in Southeast Asia: ASEAN's approach to terrorism. *The Pacific Review, 22*(2), 159–177. https://doi.org/10.1080/09512740902815300
Fravel, M. T. (2008). *Strong Borders, secure nation: Cooperation and conflict in China's territorial disputes*. Princeton University Press.
Goodhand, J. (2018). Borderlands, Brokers and Peacebuilding: War to Peace Transitions Viewedfrom the Margins. CEPA. http://www.borderlandsasia.org/uploads/1579261490_The%20Centrality%20of%20the%20Margins.pdf, [Accessed on-29 Sep 2023].
Keck, Zachary. (2014). China's Newest Maritime Dispute, *The Diplomat*, published on-28 February 2015, https://thediplomat.com/2014/03/chinas-newest-maritime-dispute/, [Accessed on-12 Jun 2023].
Mancini, F. (2013). Uncertain Borders: Territorial disputes in Asia. ISPI. https://www.ispionline.it/sites/default/files/pubblicazioni/analysis_180_2013_0.pdf
Mishra, S. (2016). The colonial origins of territorial disputes in South Asia. *The Journal ofTerritorial and MaritimeStudies, 3*(1). https://www.jstor.org/stable/26664123?seq=1
Parry, Martin., Canziani, Osvaldo., Palutikof, Jean., Linden, Paul Van der., Hanson, Clair. (2007). Climate change 2007: Impact, adaptation and vulnerability., Cambridge University Press, 32 Avenue of the Americas.
Tripathi, D., & Chaturvedi, S. (2019). South Asia: Boundaries, Borders and beyond. *Journal of Borderlands Studies., 35*(2). https://doi.org/10.1080/08865655.2019.1669483
United Nations. (2018). United Nations Must Be 'Bold and Creative' in Using Mediation to Broker Peace, Secretary-General Tells Security Council, published on-29 August 2018. https://press.un.org/en/2018/sc13475.doc.htm, [Accessed on- 29 Sep 2023].
UNODC. (2019). Transnational Organized Crime in Southeast Asia: Evolution, Growth andImpact. UNODC. https://www.unodc.org/documents/southeastasiaandpacific/Publications/2019/SEA_TOCTA_2019_web.pdf, [Accessed on- 29 sep 2023].
Wagener, M. (2011). Lessons from Preah Vihear: Thailand, Cambodia, and the nature of low-intensity border conflicts. *Journal of Current Southeast Asian Affairs*. https://doi.org/10.1177/2F186810341103000302, [Accessed on- 29 Sep 2023]
Weigand, F. (2020). *Conflict and transnational crime: Borders, Bullets & Business in Southeast Asia*. Edward Elgar.
Werleman, C. (2020). The human cost of India's Baglihar dam in disputed Kashmir. TRTWorld. https://www.trtworld.com/opinion/the-human-cost-of-india-s-baglihar-dam-in-disputed-kashmir-3879638796
Yusuf, B. (1994). The search for identity: Ethnicity, religion and political violence, UNRISD occasional paper: World summit for social development, no. 6, United Nations research Institute for Social Development (UNRISD), Geneva. https://www.econstor.eu/bitstream/10419/148780/1/862526043.pdf, [accessed on- 20 Aug 2023].

Dr. Siddhartha Sankar Manna is an Assistant Professor in the Department of Political Science at the University of Gour Banga. He was awarded with the Ph.D. from the Vidyasagar University.

The focal theme of his research is the analysis of international relations, political theory, democratic politics, and human development. He published more than twenty papers in reputed national and international journals.

Chapter 2
Environmental Geopolitics: Reflections on 1992s United Nations Conference on Environment and Development After Three Decades

Leandro Dias de Oliveira

Abstract In 1992, the United Nations Conference on Environment and Development (Rio-1992) was held. Our efforts have reinforced that the Rio-92 meant: (Acselrad in Develop Environ 56:5–8, 1993) a diplomatic adjustment in favor of the geopolitical operation of sustainable development and environmental neoliberalism; (Acselrad in Herculano, Selene C. Environment: Conceptual Issues I, 2000) the celebration of a technical arrangement for the use of nature, through technological conservation mechanisms; and (Addis Ababa Action Agenda in Addis Ababa Action Agenda of the Third International Conference on Financing for Development, 2015) the consolidation of an ideology that confuses sustainable development with saving the world. The objective of this chapter is to take stock after 30 years on the changes, impacts, successes, and errors of the sustainable development model, in its geopolitical, technical–economic, and ideological actions.

Keywords Environmental geopolitics · Earth summit · Sustainable development · Rio de Janeiro · Neoliberalism

2.1 Introduction

Environmental geopolitics is the contemporary process of dispute between the National States and large corporations for the control and appropriation of nature, understood as territory and wealth. It is a theme that has aroused interesting analysis and with great repercussions (Castree, 2003; Dalby, 2003, 2014; Pishgahifard & Alizadeh, 2007; Avendaño Castro & Aguilar Rodríguez, 2014; O'Lear, 2018, 2020; Estenssoro, 2020; Le Billon & Shykora, 2020; Basu, 2021). As the important Brazilian geographer Bertha Becker (2005) teaches us, geopolitics is "a field

---

L. D. de Oliveira (✉)
Department of Geography at Federal Rural, University of Rio de Janeiro (UFRRJ), Rio de Janeiro, Brazil
e-mail: ldiasufrrj@gmail.com

M. Mishra et al. (eds.), *Contemporary Geopolitical Alignment in the Global South*,
https://doi.org/10.1007/978-3-031-81977-3_2

of knowledge that analyzes relations between power and geographic space." But from a perspective of Critical Geopolitics, liberation from the fetishism of the State is sought (Cairo Carou, 1993) and to incorporate, among other factors, a political economy of the territory.

In this sense, the United Nations Conference on Environment and Development, held in Rio de Janeiro, Brazil, in 1992, is an important geopolitical event. Bringing together the largest number of diplomatic delegations until then, the 1992 Rio de Janeiro Conference enacted sustainable development as a goal and promoted the signing of very important environmental documents, such as Agenda 21. This meant profound and important analyses of the event (Dalby, 1996; Le Prestre, 1997; Nogué & Rufí, 2001; Castree, 2003). After all, studying a conference of this nature is justified by the understanding that events like these are capable of triggering transformations in the international order; likewise, analyzing the movements arising from these processes, and their own final documents, is undoubtedly an intellectual exercise in the field of geopolitics (Oliveira, 2019a, 2019b).

The purpose of this chapter is to briefly review the contemporary environmental issue, 30 years after the United Nations Conference on Environment and Development. After all, what are the ideas celebrated 30 years ago that are still alive? Could it be that after 30 years we have learned from the mistakes of the past? To answer these questions, a brief presentation of the Rio de Janeiro-1992 Conference will be firstly made; next, the concept of sustainable development and its relations with economic neoliberalism and the productive restructuring of capital will be discussed; finally, a brief assessment of the environmental issue currently.

2.2 The United Nations Conference on Environment and Development (Earth Summit, Rio-1992)

The United Nations Conference on Environment and Development was a large diplomatic meeting held between June 3 and 14, 1992, in the city of Rio de Janeiro. During this period, official delegations from 178 countries met at an important convention center in Barra da Tijuca called Riocentro, holding the World Summit on the Environment. 40 km away, at the Aterro do Flamengo, the Global Forum took place, an event organized by various Non-governmental Organizations (NGOs) and popular movements from all over the world, with the participation of social movements, religious leaders, various artists, intellectuals, journalists, among others.

Up to then, it was the biggest conference of all time, surpassed only in 2000 with the Millennium Summit and then with the United Nations Conference on Sustainable Development, known as Rio + 20, held in 2012. Some of the most important heads of state and government in the world attended the Earth Summit, in the city of Rio de Janeiro: George Bush, President of the United States; John Major, Prime Minister of the United Kingdom; François Mitterrand, President of France; Fidel Castro, president of Cuba; Fernando Collor de Mello, President of Brazil as host, and many others.

At Aterro do Flamengo, where the Global Forum was installed, intellectuals from different nationalities held important discussions. Academic debates with scientists and the general public, ecclesiastical and ecumenical meetings, tents on ethnic and gender issues, and meetings between leaders of indigenous and traditional communities took place at the event. There was intense participation of social movements organized in urban, agrarian, gender, ethnic, and ecological struggles and renowned or anonymous artists. Around 20,000 NGO representatives from over 170 countries worldwide participated in the Forum.

The Earth Summit was the consecration place for sustainable development (Veiga, 2005). It was a bipartite event: while the Earth Summit took place at Riocentro, which was the meeting of official delegations from participating countries where diplomatic decisions and international agreements were formed, at Aterro do Flamengo the Global Forum took place, involving NGOs and social movements, with the participation of several people interested in debating this issue. However, the Global Forum ended up being treated as a festive event where the definitive celebration of sustainable development took place as a common interest, constituting a kind of simulacrum whose intent was to build a consensus around the concept of sustainable development to be celebrated locally and globally.

If most of the signed documents convey the differences present at the conference, Agenda 21 (UNCED, 1996 [1992]) became the symbolic treaty for the realization of sustainable development as a common and universal recipe for central or peripheral countries. During the United Nations Conference on the Human Environment, held in Stockholm in 1972, there was a great confrontation between the defense of development and its freezing, largely influenced by the Limits to Growth report (Meadows et al., 1972). In Rio de Janeiro, the disagreements reverted into a coalition for sustainable development, an adaptation between economy and ecology for the central and periphery countries.

2.3 Sustainable Development, Neoliberalism, and Productive Restructuring

According to the document, Our Common Future (Brundtland, 1987), the result of a commission summoned by the UN and chaired by the Norwegian politician Gro Harlem Brundlant, sustainable development meets "the needs of the present without compromising the ability of future generations to meet their own needs." It is a comprehensive, utopian, vague, and imprecise definition. Both in Our Common Future (Brundtland, 1987) and in Agenda 21 (UNCED, 1996 [1992]), there are so many scenarios that the conception becomes malleable. Sustainable development was consolidated as the formula found to respond to the negative impacts of the industrialist conception of progress, incorporating environmental capital and abandoning the view that nature is a free good and giving it a fair price (Acselrad, 1993, 2000).

The process of modernization and capital accumulation constituted in the capitalist mode of production promoted a true ideological kidnapping of the concept of development. Environmentally, development means a rupture with the time–space of nature, which is transformed into raw material, fuel, economic resources, and present and future commodities. Sustainable development is the result of a geopolitical operation that belongs to the legacy of productive changes in the twentieth century and is contained in the gears of innovation in the world of production. Sustainable development turned out to be an economic-ecological adjustment integrated with the productive restructuring of capital.

The adoption of sustainable development did not break with the industrial, consumerist model of civilization, of constant disposal and destruction of natural wealth stigmatized as resources and commodities. This occurred because its adoption is based on technological advancement, renewing faith in the redemptive capacity of the technique (Habermas, 2014a, 2014b [1968]; Porto-Gonçalves, 2002), in addition to being carried out by the same large corporations under the renewal of their productive tools. Finally, the main interest of the model is economic: the adoption of sustainable development is a readjustment of factory production, an adjustment in the pollution load, and the inclusion of filtering mechanisms always aiming for profits.

In the very first chapter, Agenda 21 (UNCED, 1996 [1992]) states that the promotion of sustainable development should be carried out through trade liberalization. According to the same document, an open, fair, safe, non-discriminatory, and predictable multilateral trading system is the model compatible with the objectives of sustainable development. As stated, this open market will bring benefits to all trading partners, showing that sustainable development must be incorporated into the economic agenda of neoliberalism. Thus, many studies have analyzed the relationship between sustainable development and economic neoliberalism in different ways (Manteaw, 2008; Himley, 2010; Cervantes, 2013; Reid, 2013; Wanner, 2015; Adelman, 2018; Baldwin et al. 2019).

Harvey (2008) states that neoliberalism is more than an economic theory: it is a set of political actions to place the market as the main beneficiary of State actions. If in its classic form, the liberal theory assured the need for a free market provided by the non-participation of the State in the economy, in its neoliberal form this became a kind of excuse for the State to act prominently in favor of the market and big capital. The State, strictly speaking, does not hide itself, but strengthens itself in favor of capital and against the poorest layers of society, often in a violent and authoritarian way.

Dardot and Laval (2009) point out that with neoliberalism, a pattern of corporate governance emerges, through a true cult of the corporate world. This occurs with the purpose of shared management by public–private governance models under managerial formulas. In this flexible State, the corporate vocabulary is adopted by its agents, with the privatization of the construction of norms, the hybridization of public and private actions, and the selfishness of the undertaken actions, substituting the collective perspective for the corporate one and collective interests for private ones. If the State starts to be understood as a company and profit becomes its main reason, the market notions of effectiveness and performance and the economic usurpation of the term freedom end up consolidating the paradigm which causes the emptying of planning and management of public affairs and gradually transforms rights into services. Sustainable development progressively becomes corporate sustainability, that is, an environmentally correct development model for large companies.

2.4 The Legacy of the Rio de Janeiro Conference in 1992

The United Nations Conference on Environment and Development, held in Rio de Janeiro in 1992, marked an important step in international conferences with broader future consequences (Gabaldón, 1996). Rio de Janeiro was an institutional and social point of reference (Nogué & Rufí, 2001) and the culmination of the path of building sustainable development, becoming a true watershed in the perception of environmental issues (Florêncio, 2008).

The 1992 Rio de Janeiro Conference contributed to the growth of a body of principles on the environment and sustainable development (Vieira, 2002) for States, companies, and civil society groups. For Leila Ferreira (1995) Rio-92 was not the culmination of a process, but its beginning: it was a significant event in environmental geopolitics, and it remains alive in the multiplication of local agendas 21 around the world.

On the other hand, based on the works of Bellamy-Foster and Clark (2006) and Alfred Crosby (1986), one can speak of "ecological imperialism", through the hegemony of central countries in the world's environmental decisions. It would be an "*environmentalization of geopolitics*", according to the work of Nogué and Rufí (2001), involving everything from the looting of resources from certain countries by others to even the discharge of ecological waste that widens the gap between center and periphery. The geopolitics of biodiversity and sustainable development are properly inserted in the context of economic globalization and in the process of naturalization of the "commodification of nature" (Leff, 2001, 2002).

Sustainable development was consolidated after Rio-92 as an important mechanism of action for various social subjects: from States in their multiple scales to companies of different sizes, from social activists involved in NGOs, neighborhood associations, and trade unions to graduate programs at universities, from environmental education in schools, in different education segments, mainstream media special features. The ideology of sustainable development has advanced by giant leaps in the world.

Even aware of the cynicism present in many environmental measures, any ecological care is more interesting than the complete break with any concern with the defense of nature. The adoption of sustainable development did not mean a major transformation in the society-nature relationship, but it brought important advances for environmental measures to be taken by governments, companies, and civil society. Progress has been slow, but it has brought some hope for an environmentally better future. But after 30 years, the scenario is not so promising.

2.5 After 30 Years: Is There Something to Celebrate?

Twenty years after the Earth Summit, the United Nations Conference on Sustainable Development, known as Rio + 20, took place. Amidst the strengthening of questions about global warming and other environmental problems and in a scenario where the acceptance of the principles of sustainable development by the largest companies in the world did not mean structural changes, Rio + 20 did not bring great advances to the environmental agenda (Oliveira, 2014, 2019a, 2019b).

The intensive use of the term green economy at Rio + 20 represented a serious setback. After all, greening the economy is not transforming it, but practicing greenwashing, a burlesque appropriation of environmental actions with propaganda goals (Oliveira, 2022). It was also noticed the few practical effects of diplomatic mega-meetings of this nature, incapable of solving humanity's most pressing problems.

Rio + 20 confirmed the market as the main driver and funder of global environmental actions. Shortly thereafter, at the Third International Conference on Financing for Development, held in the city of Addis Ababa in 2015, it was agreed that private commercial activity would become the main driver of development. The basis is the idea that large global corporations are capable of making large investments, promoting innovation, and providing development (Addis Ababa Action Agenda, 2015).

In 2015, the Sustainable Development Summit approved the new agenda entitled "Transforming our world: the 2030 agenda for sustainable development" (General Assembly, 2015), containing the 17 Sustainable Development Goals. The proposal does not have the same ideological potency as previous documents such as Our Common Future (Brundtland, 1987) or Agenda 21 (Brundtland, 1987). By 2030, the pledge is that every country in the world will "end poverty in all its forms everywhere" (SDG 1), "achieve gender equality and empower all women and girls" (SDG 5), and "ensure access to cheap, reliable, sustainable and renewable energy for all" (SDG 7). A set of unrealizable proposals only highlights the weakness of the proposal.

In addition, the rise to power of political figures such as Donald Trump in the United States and Jair Bolsonaro in Brazil also represented a serious setback in the environmental issue. Representatives of ideas based on an overwhelming development model and their positions in the major international meetings on the matter, especially regarding climate change, were catastrophic. Devastation, pollution in its various forms, and the impacts of mining represent the barbaric legacy of denying environmental problems.

If the last decade could have already been considered environmentally lost due to the setbacks of Rio + 20, the fragility of the 17 SDGs, and the international expression of far-right leaders who despise the environment, the COVID-19 pandemic forces us to reflect on our civilization. The COVID-19 pandemic is the result of inconsequential urban growth, economic development as a hegemonic goal, forest devastation, and lack of concern for environmental issues, especially due to the ineffectiveness of sustainable development.

However, if in 2005 the oxymoron sustainable development, born in 1987, reached the age of majority (Redclift, 2005), in 2022 the term completed 35 years. Instead of a maturity expected by the time of life, the term remains immature, without personality, understanding that it is still young for further definitions. As perceived, this neutral appearance is a disguise for its true essence, which in the last decade has become more evident and public, through the conception of the green economy. 30 years after the United Nations Conference on Environment and Development, Rio-92, with the market being the master of ecological decisions, the environment and humanity are calling for help.

2.6 Final Considerations

The Brazilian geographer Santos (2002 [1996]) taught us that we live in a new historical–geographical period that he called informational technical-scientific milieu. This milieu is based on the union between technique and science, but also on technology as a norm, on the global circulation of information, and on a new interpretation of the ecological issue. The ecological question involves a new logic of subordination of nature and the deterritorialization of the ecological disaster, making the environmental crisis a result of the search for surplus value at the global level.

Rio-92 took place shortly after the fall of the Berlin Wall, in 1989, and the end of the former Soviet Union, in 1991. With the collapse of real socialism, an era of robust commercial exchanges happened through customs barriers elimination and a neoliberal political, economic, and social order. Globalization was the promise of a world built through dialogue without major enemies and without military and war actions. The world of invasion threats and military confrontations would become a historical record of the Cold War period.

Thirty years after Rio-92, we live in times of war, with the invasion of Ukraine by Russian troops, leaving a trail of more than 100,000 Moscow soldiers dead or wounded, as well as Ukraine probably suffering a similar number of casualties (Cooper, 2022). In the same way, thousands of people still die every day due to the effects of the COVID-19 pandemic, in an unprecedented medical and health disaster in history. The belief in neoliberalism, in the green economy, and in the market as a promoter of development is shaken for most people in the world. But let's not wait for neoliberal economists to repent and question themselves in front of the mirror: for such economists, the mistakes of neoliberalism are always because there is still too much State and a lack of market and freedom of trade. Meanwhile, the environment suffers from the advances of unsustainable development.

In the United States, democracy was tested with the election and inauguration of Joe Biden in 2021, when Donald Trump voters invaded the Capitol in a frustrated attempt to avoid the new president's inauguration. In a similar way, but much more seriously, in Brazil, Jair Bolsonaro's voters invaded in January 2023 and destroyed the headquarters of the Executive, Judiciary, and Legislative powers, in an attempt to overthrow the elected president, Luís Inácio Lula da Silva, and with the connivance of those who should protect public property—military, police, national guard. The promised post-Cold War democracy has suffered heavily from the emergence of a violent, coup-mongering, and persecutory far right.

Therefore, more than celebrating, we must be attentive to the challenges posed in relation to the environment, but also to politics, the economy, social issues, culture, and democracy itself. Unfortunately, the world remains very sexist, and racist, with religious prejudices and little ecological awareness. On the subject of this chapter, it is urgent to deal seriously with the environmental issues: climate change, forest devastation, pollution in its various forms, loss of biodiversity, and a multitude of ecological problems that continue to put our future at risk.

Acknowledgements This analysis is an integral part of the Political Ecology project, territorial-productive restructuring, and sustainable development: implications in the metropolitan peripheries, awarded with the Research Productivity Scholarship, Level 2, National Council for Scientific and Technological Development (CNPq), and our State Young Scientist Scholarship, from the Carlos Chagas Filho Foundation for Research Support in the State of Rio de Janeiro (FAPERJ). We appreciate the support of these funding agencies in the development of our research.

References

Acselrad, H. (1993). Sustainable development: The struggle for a concept. Proposal: experiences in popular education. *Development and Environment*, 5–8. Rio de Janeiro, FASE, No. 56, Year XVII.

Acselrad, H. (2000). Sustainability, space and time. In *Herculano, Selene C. Environment: Conceptual Issues I*. Niterói, PGCA—Riocor.

Addis Ababa Action Agenda. (2015). *Addis Ababa Action Agenda of the third international conference on financing for development*. Addis Ababa, Ethiopia, 13–16 July 2015, General Assembly, Resolution 69/313, United Nations, New York, 27 July 2015.

Adelman, S. (2018). The sustainable development goals, anthropocentrism and neoliberalism. In D. French, L. J. Kotzé (Eds.), *Sustainable development goals: Law, theory and implementation*. Edward Elgar Publishing;

Avendaño Castro, W. R., & Aguilar Rodríguez, D. E. (2014). Geopolitics and the environment: A look at the problem of environmental displacement. *Research and Development, 22*(2), 283–308.

Baldwin, C., Marshall, G., Ross, H., Cavaye, J., Stephenson, J., Carter, L., & Syme, G. (2019). Hybrid neoliberalism: Implications for sustainable development. *Society & Natural Resources, 32*(5), 566–587.

Basu, J. (2021). Geopolitics in environmental discourse. In *Environmental management: Issues and concerns in developing countries* (pp. 323–330). Springer.

Becker, B. K. (2005). Geopolitics of the Amazon. *Advanced Studies* [online], *19*(53), 71–86. https://doi.org/10.1590/S0103-40142005000100005

Bellamy-Foster, J., & Clark, B. (2006). Ecological imperialism: the curse of capitalism. In L. Panitch, C. Leys (Eds.), *The new imperial challenge.* Buenos Aires, Consejo Latinoamericano de Ciencias Sociales—CLACSO.
Brundtland, G. H. (Ed.). (1987). Oxford University Press.
Cairo Carou, H. (1993). Elements for a critical geopolitics: tradition and change in a cursed discipline. *Ería*, (32), 195–213. https://doi.org/10.17811/er.0.1993.195-213
Castree, N. (2003). The geopolitics of nature. In J. Agnew, K. Mitchell, G. Toal (Eds.), *A companion to political geography*. Blackwell Publishing Ltda.
Cervantes, J. (2013). Ideology, neoliberalism and sustainable development. *Human Geographies – Journal of Studies and Research in Human Geography*, *7*(2), 25–34.
Cooper, H. (2022). Russia and Ukraine each have suffered over 100,000 casualties, the top U.S. general says. The New York Times, 10/11/2022. https://www.nytimes.com/2022/11/10/world/europe/ukraine-russia-war-casualties-deaths.html
Crosby, A. W. (1986). *Ecological imperialism: The biological expansion of Europe, 900–1900.* Cambridge University Press/The Press Syndicate of the University of Cambridge.
Dalby, S. (2003). Environmental geopolitics: Nature, culture, urbanity. In K. Anderson, M. Domosh, S. Pile, N. Thrift (Eds.), *Handbook of cultural geography* (pp. 595–612). Sage.
Dalby, S. (1996). Reading Rio, writing the world: The New York Times and the 'Earth Summit.' *Political Geography, 15*(6–7), 593–613.
Dalby, S. (2014). Environmental geopolitics in the twenty-first century. *Alternatives, 39*(1), 3–16.
Dardot, P., & Laval, P. (2009). *The new reason of the world* (p. 504). Éditions La Découverte.
de Oliveira, L. D. (2014). Rio + 20: Reflections on geopolitics and ideology. *Space and Economy: Brazilian Journal of Economic Geography*, *2*(4). http://journals.openedition.org/espacoeconomia/854
de Oliveira, L. D. (2019b). Unlimited development? The contemporary environmental issue from the perspective of economic geography. In F. J. G. Oliveira, L. D. Oliveira, R. H. Tunis, R. M. Pessanha (Eds.), Space *and economy: Economic geography and political economy.* Rio de Janeiro, Consequence.
de Oliveira, L. D. (2019a). *Environmental geopolitics: The ideological construction of Sustainable Development (1945–1992).* Rio de Janeiro, Autografia. https://www.researchgate.net/publication/344349857_Geopolitica_Ambiental_A_construcao_ideologica_do_Desenvolvimento_Sustentavel_1945-1992
de Oliveira, L. D. (2022). *2022 and the jubilee of major environmental conferences.* Le Monde Diplomatique Brasil, 06/24/2022. https://diplomatique.org.br/2022-e-o-jubileu-das-grandes-conferencias-ambientais/
Estenssoro, F. (2020). Challenges of Latin America in the global environmental geopolitics of the twenty-first century. In *Latin America in Times of global environmental change* (pp. 15–29). Springer.
Ferreira, L. C. (1995). Environmentalism after Eco-92. *Ideas: Journal of the Institute of Philosophy and Human Sciences.* Campinas, year 2, no. 1.
Florêncio, S. A. L. (2008). Brazil and the challenge of developing the United Nations. In *Brazil and the UN (United Nations).* Brasília, Alexandre de Gusmão Foundation.
Gabaldón, A. J. (1996). D*ialectic of sustainable development: A Latin American perspective.* Polar Foundation.
General Assembly of United Nations (A/RES/70/1). (2015). Transforming our world: the 2030 Agenda for Sustainable Development, 25 September 2015.
Habermas, J. (2014 [1968]). Technique and science as "ideology". [Original title in German: Technick und Wissenschaft als "Ideologie"]. São Paulo, EdUNESP.
Habermas, J. (2014 [1968]). Técnica e ciência como "ideologia". [Original title in German: Technick und Wissenschaft als "Ideologie"]. São Paulo, EdUNESP.
Harvey, D. (2008 [2005]). *A brief history of neoliberalism.* Loyola.
Himley, M. (2010). Global mining and the uneasy neoliberalization of sustainable development. *Sustainability, 2*(10), 3270–3290.

Le Prestre, P. (1997). Écopolitique internationale. Québec, Guérin.
Le Billon, P., & Shykora, L. (2020). *Conflicts, commodities and the environmental geopolitics of supply chains*. Edward Elgar Publishing.
Leff, E. (2001). The geopolitics of biodiversity and sustainable development. Paper presented to the IIa Conference of CLACSO (Latin American Council of Social Sciences). Guadalajara Mexico. http://bibliotecavirtual.clacso.org.ar/ar/libros/osal/osal17/dleff.pdf
Leff, E. (2002). The geopolitics of biodiversity and sustainable development: Economization of the world, environmental rationality and social reappropriation of nature. In A. E. Ceceña, E. Sader (Eds.), *The infinite war—Hegemony and world terror* (pp. 191–216). Latin American Council of Social Sciences. http://www.itescham.com/Syllabus/Doctos/r418.PDF
Manteaw, B. O. (2008). When businesses go to school: Neoliberalism and education for sustainable development. *Journal of Education for Sustainable Development, 2*(2), 119–126. https://doi.org/10.1177/097340820800200209
Meadows, D. H., Meadows, D. L., Randers, J., & Behrens III, W. W. (1972). *The limits to growth*. Potomac Associates.
Nogué, J., & Rufí, J. V. (2001). *Geopolítica, identidad y globalización*. Ariel.
O'Lear, S. (2018). *Environmental geopolitics*. Rowman & Littlefield.
O'Lear, S. (Ed.). (2020). *A research agenda for environmental geopolitics*. Elgar Research Agendas.
Pishgahifard, Z., & Alizadeh, O. (2007). The importance of Environmental Geopolitics in today's world. *Territory, 4*(13), 22–28.
Porto-Gonçalves, C. W. (2002). The invention of new geographies. In *Postgraduate program in geography*. Territory Territories. Niterói, UFF/AGB.
Redclift, M. (2005). Sustainable development (1987–2005): An oxymoron comes of age. *Sustainable Development, 13*(4), 212–227.
Reid, J. (2013). Interrogating the neoliberal biopolitics of the sustainable development-resilience nexus. *International Political Sociology, 7*(4), 353–367.
Santos, M. (2002 [1996]). The nature of space. *Technique and time. Reason and emotion*. EDUSP.
UNCED. (1996 [1992]). United Nations Conference on Environment and Development. Agenda 21. Brasília: Federal Senate, Subsecretariat for Technical Editions.
Veiga, J. E. (2005). *Sustainable development: The challenge of the 21st Century*. Garamond.
Vieira, S. C. (2002). The construction of the concept of sustainable development. In D. P. R. Fonseca, J. C. Siqueira (Eds.) *Environment, culture and development*. Rio de Janeiro, Sette Letras/Historia y Vida.
Wanner, T. (2015). The new 'passive revolution' of the green economy and growth discourse: Maintaining the 'sustainable development' of neoliberal capitalism. *New Political Economy, 20*(1), 21–41.

Dr. Leandro Dias de Oliveira has graduated in Geography from the State University of Rio de Janeiro (UERJ), Master in Geography from the State University of Rio de Janeiro (UERJ), Doctor of Geography from the State University of Campinas (Unicamp) and Post-Doctor in Public Policy and Human Training from the University of the State of Rio de Janeiro (UERJ). Associate Professor of the Department of Geography at UFRRJ—Federal Rural University of Rio de Janeiro (DGG-IGEO), headquarters campus, and professor of the permanent staff of the Graduate Program in Geography (PPGGEO-UFRRJ) and the Graduate Program Interdisciplinary Graduation in Digital Humanities (PPGIHD-UFRRJ). He integrates and coordinates the Laboratory of Economic Geography, Policy, and Planning (LAGEP). Research Productivity Scholarship, Level 2, National Council for Scientific and Technological Development (CNPq), and our State Young Scientist Scholarship, from the Carlos Chagas Filho Foundation for Research Support in the State of Rio de Janeiro (FAPERJ).

Chapter 3
Peace in Practice and Peace in SDG Framework: Revisiting the Path of Nonviolence

Mukunda Mishra

Abstract Nonviolence is arguably the finest 'weapon' to achieve 'positive peace.' India, for a long, has been the gene center of this idea—through the *hymns* of Vedas, *mantras* of Buddhists, and many other religious ideas prevailing for thousands of years over this vast territory. All these have shaped the ideas of *'ahimsa'* (non-violence) and *'satyagraha'* (truth force) of Gandhi as a path of social and political change for sustainable peace. The concepts of *ahimsa* and *satyagraha* are the normative principles of peace-building. The actor-oriented models of peace explain peace to have prevailed through the balance of powers. While there are disagreements about what is peace, and how peace may be achieved, the positive role of peace on development has unanimously been accepted by all. United Nations' early initiative of the 'right to peace' was transformed into the shape of the 'culture of peace' which further evolved to set the background idea of SDG-16 in the Sustainable Development Goals (SDG) framework, adopted by the UN in 2015 to count achieving peace as a goal of sustainable development. This chapter presents a comprehensive review of the context of the inclusion of 'peace' in the SDG framework and proposes the practice of nonviolence as the pathway to achieving peace and sustainability in the long run.

Keywords Geopolitics · Non-violence · Right to peace · Culture of peace · Positive peace · Powers · SDG-16

M. Mishra (✉)
Dr. Meghnad Saha College, Itahar, West Bengal, India
e-mail: mukundamishra01@gmail.com

M. Mishra et al. (eds.), *Contemporary Geopolitical Alignment in the Global South*,
https://doi.org/10.1007/978-3-031-81977-3_3

3.1 Towards 'Positive Peace'

Human civilization since its known history has been prioritizing peace over conflict under any circumstances as annotated by Erasmus[1] that 'the most disadvantageous peace is better than the most just war.' There are disagreements among scholars about the definition of peace but all commonly agree that a state of peacefulness has a greater likelihood of producing 'good' for humankind in any form. This chapter starts with an explanation of how our intuitive expectation of peace differs from the 'peace' that exists in reality; while the former is driven by the conceptual thoughts of individuals with highly theorized connaissance, the latter is linked with the socio-political equations across and within the state.

A casual idea of 'peace' is something that is opposed to 'war' in any form—international, civil, or revolutionary. Hence, by dictionary definition, 'peace' lies at antipole to 'war' as the latter, by all means, sets a cognitive image of destruction, humilities, and hostilities. One classical pacifist would, however, count 'violence' as well in this perspective of defining peace because the term 'violence' additionally covers violent acts that are not legally defined as war. So a peaceful condition (maybe the state of mind or a territory) in a more strict sense is free from violence of all kinds. Due to the invocation of different international laws, 'wars' in a 'legal sense' are counted less in numbers than the actual cost of lives lost or hostilities to personal, social, and cultural values. From that standpoint, peace is conceptualized that it is only those human relations that exclude violence in any form - personal, organized, or collective (Rummel, 1981). However, attaining peace through the absence of war, violence, and antagonistic conflict, such as manifesting threats and accusations, hostile quarrels, angry boycotts, and riotous demonstrations is not all about peace in the truest sense.

Interestingly, until Galtung (1964) presented peace in a new form in his editorial text of the founding edition of the Journal of Peace Research, the social theorists' position about peace was not far different from the ideas that 'peace means the halt of war.' Peace from psychological, social, and political senses has long been superimposed to consequent into the actor-oriented explanation of peace. Galtung (1985) founded the cornerstone of the structure-oriented model of peace, which is unlike the previous actor-oriented peace model, categorizes peace into two types—'positive peace' and 'negative peace.' He has the initial conceptualization of his idea of peace from the basics of human health science. He explains that 'good health' may be meant as merely the absence of illness (which is a pessimistic idea) or maybe something more, something positive to body and mind that could resist future diseases (which is an optimistic idea). Similarly, 'negative peace' is the prevailing scenario that confirms only no war is happening, no conflict is in place and no violence is

[1] Dutch humanist Desiderius Erasmus Roterodamus authored an annotated collection of Greek and Latin proverbs, compiled during the Renaissance which is entited as 'Adagia.' The first edition, titled 'Collectanea Adagiorum,' was published in Paris in 1500, which was a collection of aproximately eight hundred entries. By 1508, Erasmus had expanded the collection to over 3000 items, some of which were brief essays on political and moral topics.

in the news, however, it does neither guarantee how 'peacefully' that 'peace' has been achieved nor forecast how long it will sustain or how all citizens are benefited from this peace process quite equally. Whereas, positive peace is the effort to integrate human society that uproots the cause of those wars, conflicts, and violence. However, 'positive' peace would hypothetically not emerge unless the 'negative' peace appears. Positive peace in society would be facilitated through proper policies enabling improved human understanding through barrier-free communication and peace education, which is unlike the ideas of peace through arms control or multilateralism (Schmid, 1968; Galtung, 1969; Wiberg, 1981; Grewal, 2003).

3.2 Peace Through the Balance of Power

> "For the sake of greedy desire, kings and kings are in conflict,
>
> So are monks and monks, people and people, regions and regions, states and states"
>
> (The Middle Length Discourses of the Buddha, Taisho 28: 1547)
>
> "If we win, we incur resentment toward ourselves.
>
> If we lose, our self-esteem is hurt."
>
> (Dhammapada, Taisho 4: 210)

The theoretical cognizance of establishing peace has long been articulated in the form of 'peace' through the 'delicate balance of powers' which originated around the sixteenth century West and has evolved over time. However, the initial conceptualization of the balancing of power through the spontaneously raised order (because the balance of power was once believed to be achieved by states pursuing their self-interests) has largely been replaced by the ideas of the British School which considers the 'balance of power' as a 'manually operated' system:

> [...] the process of equilibrium is a function of human contrivance, with emphasis on the skill of diplomats and statesmen, a sense of community of nations, of shared responsibility, and a desire and need to preserve the balance of power system.
>
> (Schweller, 2016: 4)

The adherents of this idea speak in favor of robust defense arrangements to deter aggression and normalize the possibility of conflicts. However, in an ambiance of the (narrow) competitions between states to enhance respective power quotients would have no finite limit. Furthermore, the possibility of achieving equal power is impossible in a practical sense. Straightforwardly, the states' funds, energy, and attention have every possibility to be concentrated on military matters, getting shifted away from the basic peace-building mechanism. Priorities set by the authorities state, although, not all but those who desire to control others, to ensure an advantageous position in unforeseen 'war,' if that becomes 'unavoidable.' Hence, under the purview of the 'bancing between powers,' war is supposed to be a legitimate tool of statecraft.

> Balancing behaviors are preparations for war, not peace. If a major power war eventually breaks out, as it did in 1914 and 1939, there is no reason to conclude that the balance of power failed to operate properly. Quite the opposite: balance of power requires that "war must be a legitimate tool of statecraft." The outbreak of war, therefore, does not disconfirm but, in most cases, supports the theory.
>
> (Schweller, 2016: 6; *See also* Jervis, 1986: 58-79)

At the core of its theoretical architecture, the 'balance of power' keeps the possibility open that the states would adjust their differences in the battleground (Lasswell, 1965). And, worryingly, the possibility of conflicts, which would not likely be 'non-violent,' shapes the diplomatic and strategic behaviors of states and influences the geopolitical equation of the system—regionally, and even, globally in certain cases. Schweller (2016) opines that not only just the prospect of war triggered the equation of past multipolar and bipolar systems, but the anticipation that powerful states would if given the right odds, carry out territorial conquests at each other's expense shaped and shoved actors in ways complied the predictions of realism's keystone theory. Power has historically been ambitious for 'more power' making the peace propaganda at the rear of states' interest.

3.3 Right to Peace

It is appropriate to begin this section with two different resolutions adopted by the United Nations General Assembly. The former one is the '*Declaration and Programme of Action on a Culture of Peace*' which was adopted by the UN General Assembly on 13 September 1999 (United Nations, 1999) after a prolonged ten months of negotiations to prepare for the International Year for the Culture of Peace. More than fifteen years later the General Assembly on 25 September 2015 adopted the resolution on '*Transforming our world: the 2030 Agenda for Sustainable Development*' which sets the agreement by world nations on 17 Sustainable Development Goals as the central normative framework for sustainable development worldwide (United Nations, 2015a). The mentioning of the two UN resolutions does not imply my intention to make the readers aware of the chronology that conceptualizing a '*culture of peace*' appears more than a decade prior to the conceptualization of '*sustainable development goals.*' Rather, we need to understand how the efforts of codifying 'peace' take a different path than that of the matters of codifying '*sustainable development*' while '*human rights,*' '*development*' and '*peace*' are considered as three conditions that cannot exist in isolation from one another.

An early attempt by UNESCO to codify the right to peace was led by Federico Mayor, the then Director-General. In January 1997, Mayor prepared a declaration on the human right to peace, emphasizing 'peace' as a prerequisite for the exercise of all human rights and duties. This declaration elaborated that the right to live in peace should be added to the list of already recognized human rights. When this declaration was presented to the Secretary-General of the UN, the Ministers of Foreign Affairs, and Ministers of Education of the Member States, NGOs, human rights centers, and academic and educational institutions, it aroused great interest, recognition, and support from all sects (Roche, 2003). Next a meeting of experts on the human right to peace from 23–25 February 1997 which was organized by the University of Las Palmas, the Tricontinental Institute of Parliamentary Democracy and Human Rights, and UNESCO, with the support of the Government of the Canary Islands, in Las Palmas (Spain) recognized the intimate linkage between human rights and peace. With the scholarship of the Norwegian Institute of Human Rights, the draft of the declaration got its final shape. The Report by the Director-General on the Human Right to Peace in the UNESCO General Conference, 29th Session at Paris has vividly mentioned the matter in its Sect. III, Serial No. 13 and 14 (UNESCO, 1997):

> 13. On the initiative of Dr A. Eide, Director of the Norwegian Institute of Human Rights, a meeting on the human right to peace was held in Oslo from 6 to 8 June 1997. The Director of the Institute chaired the debates. […]
>
> 14. The main objective of the meeting was to prepare a Draft Declaration on the Human Right to Peace. Participants once again agreed that the preparation of such a declaration in the new circumstances created by the fall of the Berlin Wall and in an international context of violence and internal conflicts was of utmost importance. […]

This draft Declaration (Table 3.1) focused on broadening the human dimension of peace and dividing the right into three interrelated components. While draft Article 1 ascribed peace as a right inherent to humanity and considered war, violence of any kind, and insecurity as intrinsically incompatible with the human right to peace, the draft article-2 elaborates on the act of maintenance and construction of peace as the duty for all global actors to prevent the manifestations of any armed conflicts and violence. The draft article 3 earmarked the 'Culture of Peace' as a 'strategy' of engraving peace in the minds of people through education, communication, ethical practices, and upholding democratic ideas (Roche, 2003).

The Director-General of UNESCO wrote in July 1997 to the heads of all Member States, presenting his ideas on the human right to peace as contained in the Declaration on the Human Right to Peace along with informing in detail them about the development of this idea at the meetings held in Las Palmas de Gran Canaria and Oslo (Norway) and to present briefly the normative background in order to elicit the opinions of the member states. Reportedly, 42 Member States replied to the Director-General's letter where most of the then 'great' powers (e.g., USA, Russia, or, UK) and 'potential' powers (e.g., China or India) kept themselves non-responded. Twenty-eight of the 42 replies received voiced support. Another ten Member States suggested changes or proposed amendments to the Draft Declaration upon agreed on principle with the initiative. However, four replying Member States expressed reservations regarding the possible adoption by UNESCO of the Draft Declaration. They were of the view that UNESCO should focus its efforts on its own fields of competence rather than on a declaration relating to human rights (UNESCO, 1997). This 'great' initiative then went under the 'skeptical' scanners of the Members of the Executive Board. Questions were raised by 25 members of the Executive Board in their statements to the debate in the plenary of the 152nd session, as reported in the Report of the Director General, UNESCO (UNESCO, 1997):

Table 3.1 Article 1, 2 and 3 in the draft version of oslo declaration on the human right to peace

Draft article 1 Peace as a human right	Draft article 2 Peace as a duty	Draft article 3 Peace through the culture of peace
• Every human being has the right to peace, which is inherent in the dignity of the human person. War and all other armed conflicts, violence in all its forms and whatever its origin, and insecurity also, are intrinsically incompatible with the human right to peace; • The human right to peace must be guaranteed, respected and implemented without any discrimination in either internal or international contexts by all States and other members of the international community	• Every human being, all states and other members of the international community and all peoples have the duty to contribute to the maintenance and construction of peace, and to the prevention of armed conflicts and of violence in all its forms. It is incumbent upon them notably to favour disarmament and to oppose by all legitimate means acts of aggression and systematic, massive and flagrant violations of human rights which constitute a threat to peace • As inequalities, exclusion and poverty can result in the disruption of peace both at the international level and internally, it is the duty of states to promote and encourage social justice both on their own territory and at the international level, in particular through an appropriate policy aimed at sustainable human development	• The culture of peace, whose aim is to build the defenses of peace in the minds of human beings every day through education, science and communication, must constitute the means of achieving the global implementation of the human right to peace; • The culture of peace requires recognition and respect for - and the daily practice of—a set of ethical values and democratic ideals which are based on the intellectual and moral solidarity of humanity

Source Adopted from Roche (2003a)

> Some Members of the Board supported the idea of such a declaration, underlining that peace is a major precondition for the implementation of all human rights, and that human right to peace is really one of the most fundamental rights of a human being, other members expressed doubts on the advisability of elaboration by UNESCO of this kind of declaration. The Executive Board agreed with the proposal of the Director-General to include the item concerning the human right to peace in the agenda of the General Conference, in order to discuss the Report with the Oslo Draft Declaration and resolve on further steps.

UNESCO's General Conference on 06 November 1997 experienced a critical debate on the Oslo Draft Declaration which is summarized by Roche (2003a, 119) as:

> One European country after another either attacked or expressed reservations about the right to peace and accused Mayor of over-stepping his mandate. Countries from the South struck back, accusing the North of wanting to protect their arms industries. At the end, Paraguay

stated, "This rich discussion shows that the culture of peace is the central issue … and that the Human Right to Peace is needed for individuals and states." Noting that the debate split North and South, Paraguay added, "Perhaps peace is a greater concern in the South where scarce resources are being diverted to war."

3.4 Culture of Peace

The outcome of all these arrays of discussion was that UNESCO's efforts of establishing the 'right to peace' decomposed into a more 'digestive-for-all' form—the 'culture of peace' which is, undoubtedly, a modest approach to 'life that seeks to transform the cultural roots of war and violence into a culture where dialogue, respect, and fairness govern social relations' (Roche, 2003a). The culture of peace is believed to be the precondition for sustainable development as the practice of setting human cognition from 'hostility' to 'solidarity' is the cornerstone of peacemaking. Through many other channels, the culture of peace is (theoretically) effective for the globe by establishing democracy, halting violence in all its forms, offering free-flow information, reducing inequalities in many forms and so on which as a cumulative form ensures the development to happen in sustainable ways (Fig. 3.1). However the gap between theoretical expectations and realities keeps on widening. The fatalities due to battles worldwide keep on rising gradually keeping the prospect of sustainable development at stake. And, worryingly a growing number of fatalities in internationalized intra-state conflicts (Fig. 3.2a) sought more cooperation between states. Paralley, the Middle-Eastern, Asian, and African nations have been the worst sufferers during the last two decades (Fig. 3.2b) which needs careful discussions and resolutions at global forums.

While the new millennium started with the prospect of globalization to bring the distant world nations closure virtually, it kept the discussions alive that 'peace' in the truest sense had still not been established as 'violence' had yet not been eradicated and 'wars and conflicts' had not been stopped through effective dialogues. With this backdrop, world leaders met at the UN Headquarters, in New York in September 2000 and adopted the UN Millennium Declaration (United Nations, 2000). This declaration, among many other goals, focused more on reducing extreme poverty which was prevailing at a worrying level in many states of South and Southeast Asia, Sub-Saharan Africa, and Latin America. The commitment of world nations to a new global solidarity, cooperation, and partnership to achieve a series of time-bound targets by 2015 set the cornerstone of the Millennium Development Goals (MDGs), a coarse outline of which was presented by BAN Ki-moon, the then Secretary-General of the United Nations (United Nations 2002):

"Eradicating extreme poverty continues to be one of the main challenges of our time, and is a major concern of the international community. Ending this scourge will require the combined efforts of all, governments, civil society organizations and the private sector, in the context of a stronger and more effective global partnership for development. The Millennium Development Goals set timebound targets, by which progress in reducing income poverty, hunger, disease, lack of adequate shelter and exclusion — while promoting gender equality,

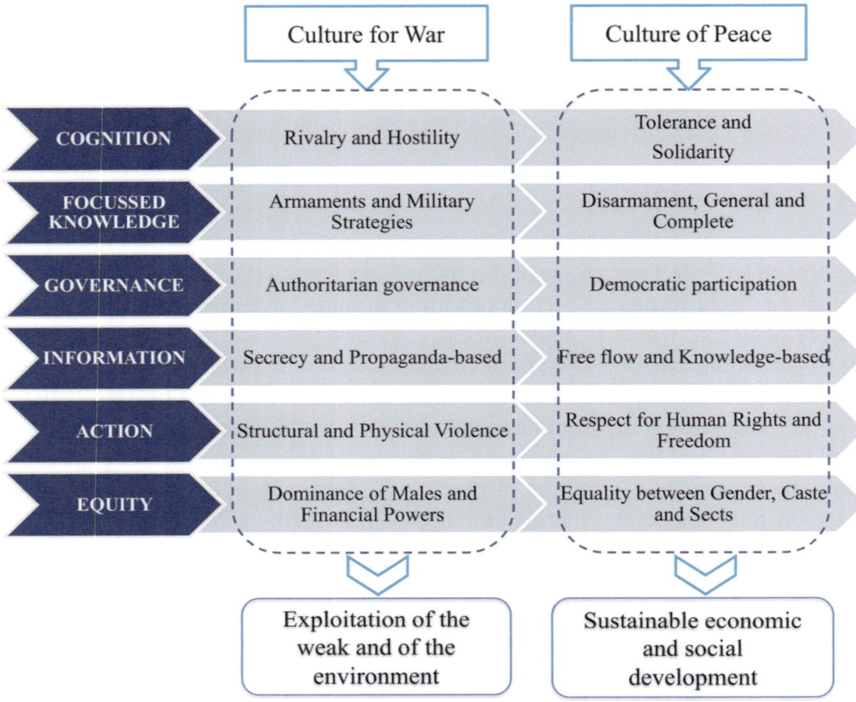

Fig. 3.1 Transformation from the culture of war to the culture of peace is expected to catalyze sustainable development through multiple channels. *Source* Prepared by the author after modification of the ideas of Roche (2003)

> health, education and environmental sustainability — can be measured. They also embody basic human rights — the rights of each person on the planet to health, education, shelter and security. "

Interestingly, 1990s dialogues of the 'right to peace' or its soft version of 'culture of peace' stayed untouched mostly in the discussions of MDGs. MDG-8 entitled 'Global Partnership for Development' had the scope to include the 'peace' propaganda as a 'target'. However, in its finalized version, none of the targets were set regarding the preservation of 'peace' in every corner of the world through effective intervention to ensure 'zero' conflict, 'zero' violence, and 'zero' war. The targets under MDG-8 were set to ensure non-discrimination of trading and financial systems, look after the special needs of least-developed countries and land-locked nations, deal with the debt-trapped situation of developing states, enhance the accessibility of developing states to essential drugs and bring the benefits of modern information and communication technology to every corner of the globe (Table 3.2). The attainment of a 'peaceful state of the world' remains uncounted in the MDGs and the UN shifted to the framework of Sustainable Development Goals (SDGs) with the commitment to achieving them by 2030.

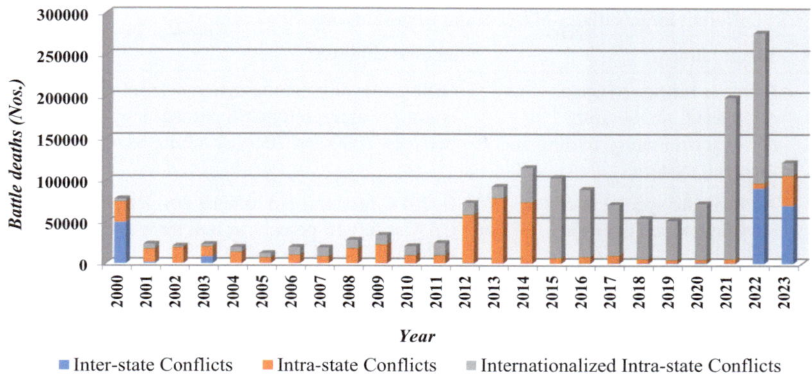

[Note: 'Inter-state Conflict' means both sides are states in the *Gleditsch and Ward membership syste*m; 'Intra-state Conflict' means one side is always a government where another side is always one or more rebel groups; there is no involvement of foreign governments with troops; and, 'Internationalized Intra-state Conflict' means one side is always a government and another side is always one or more rebel groups, with there is involvement of foreign governments with troops]

(a)

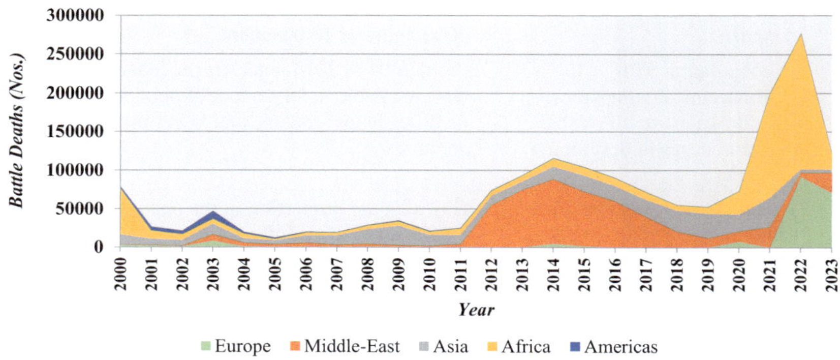

[Note: 'Americas' = Territories with *Gleditsch and Ward No. (i.e., GW No.)* 2-199; 'Europe' = *GW No.* 200-399; 'Africa'= *GW No.* 400-626; 'Middle East' = *GW No.* 630-699; 'Asia' = *GW No.* 700-999]

(b)

Fig. 3.2 No. of deaths occurred in battles **a** type of conflict wise and **b** region-wise during 2000–2023. *Source* Prepared by the author based on Uppsala Conflict Data Program UCDP, 2024 and Davies et al. (2024)

3.5 Peace as an SDG

After a complete 'no show' in MDG measures 'peace' again appeared as 'crucial' for sustainable development in the SDG framework. The current global indicator framework of SDG was adopted by the General Assembly of the United Nations on 6 July 2017 (*United Nations,* 2017). SDG framework is equipped with a robust follow-up and review mechanism for the implementation of the 2030 Agenda for Sustainable Development which has made it different immensely from the MDG framework. A solid framework of indicators and statistical data for monitoring the progress globally, informed policy, and ensured accountability of all stakeholders

Table 3.2 Different targets under MDG-8 and important outcomes achieved

Target	Target name	Outcome achieved
8.A	Develop further an open, rule-based, predictable, non-discriminatory trading and financial system	• Official development assistance from developed countries increased by 66 percent in real terms between 2000 and 2014, reaching $135.2 billion
8.B	Address the special needs of least developed countries	• In 2014, bilateral aid to least developed countries (LDCs) fell 16 percent in real terms, reaching $25 billion • 79 percent of imports from developing countries enter developed countries duty-free
8.C	Address the special needs of landlocked developing countries and small island developing states	• Projections indicate that a 2.5 percent increase in country programmable aid in 2015, mainly through disbursements by multilateral agencies, will most benefit least developed and other low-income countries
8.D	Deal comprehensively with the debt problems of developing countries	• In 2013, the debt burden of developing countries was 3.1 percent, a major improvement over the 2000 figure of 12.0 percent
8.E	In cooperation with pharmaceutical companies, provide access to affordable essential drugs in developing countries	• From 2007 to 2014, on average, generic medicines were available in 58 percent of public health facilities in low-income and lower-middle-income countries
8.F	In cooperation with the private sector, make available benefits of new technologies, especially information and communications	• Globally, the proportion of the population covered by a 2G mobile-cellular network grew from 58 percent in 2001 to 95 percent in 2015. • Internet use penetration has grown from just over 6 percent of the world's population in 2000 to 43 percent in 2015. 3.2 billion people are linked to a global network of content and applications

Source Collected from United Nations (2015b)

have helped the SDG framework to touch targetted milestones for many targets, and obviously, some more are yet to be fulfilled. However, the re-instilling of peace, and acknowledging its role in the sustainable development framework makes the SDG framework a globally valid measure of sustainable development ever in history. The Report of the Secretary-General in the UN General Assembly on 02 May 2024 has a reflection of the same (United Nations 2024: 22):

> Around the world, we need peace in every sense. Yet the road to global peace and security has become even more complex. Global forced displacement is at an all-time high, exceeding 110 million people forced to flee from persecution, conflict, violence and human rights violations as of May 2023. Between 2022 and 2023, civilian casualties experienced the highest spike since the adoption of the 2030 Agenda. Persistent threats to human security across the globe underscore the breakdown of peaceful and inclusive societies, crucial for sustainable development and the need to restore trust and strengthen and renew global peace and security frameworks to heed the desperate pleas for peace from countless voices worldwide.

The SDG-16, entitled 'Promote peaceful and inclusive societies for sustainable development, provide access to justice for all and build effective, accountable and inclusive institutions at all levels' is set with twelve targets which is measured by a total of 24 indicators (Table 3.3). As a whole, SDG-16 provides a comprehensive measurement of the achievement (or failure, whatever may be as actual) toward attaining peace and solidarity. It not only measures a 'part' of the contemporary development agenda but rather assesses, in the truest sense how far sustainable would the development be for a particular nation or region. Peace is the crucial factor that guarantees the longevity of the development process in any form.

The broader human rights framework underpins the core architecture of SDG-16 which appears to promote societies that respect and uphold individual rights; also the right to privacy, freedom of expression, and access to information. The UN[2] considers peace as a fundamental precondition for social and economic development because societies without peace have a higher likelihood of experiencing violence, and instability that consequent in a halt to the development process. SDG-16 seeks equal access to justice. It ensures the protection of the rights of individuals and resolves disputes, particularly caring for vulnerable populations not to be marginalized or mistreated.

A critical superimposition of the geographic space and the political space has critically evolved into several forms to reach their present shape (or sense) of geopolitics. The evolution of dynamic spatial sense over time on one hand and the changing political (let's say, merged with economic agenda) throughout the globe has made the forms, patterns, and process of geopolitics a large vista of critical debates that encompasses a careful connotation of social-scientists. Ratzel's organic state theory, proposed in 1897, molded under social Darwinism as an intellectual movement between the late 19th and early 20th centuries to further evolve through the conceptualization of geopolitics as a discipline to addressing political power linked to geographical space, ultimately forming the theoretical foundation of critical geopolitics where the influence of the post-structural approach is a great concern to shape it to what it is today. If the world needs to see the role of power in attaining peace at different corners of the globe that could set a comparatively newer interpretation of geopolitical equations that how the process of peacebuilding catalyzes sustainable development. While a group of present scholarship efforts to measure the comparative power orders of the world states (e.g., The Great Power Index 2024 by Ray Dalio; or, the Asia Power Index by Lowy Institute since 2018) which appears to be a newer turn in the discussion of geopolitics during the last two decades, the same is applied for peace (e.g., Global Peace Index 2024 by the Institute for Economics & Peace). Such measurements are comprehensive to provide a global scenario of powers or peace separately. Parallely, the SDG-16 (along with other SDGs) measurement framework works to understand how far the states are heading ahead to achieve Sustainable Development. However, all these measurement techniques do not complement each other which is a challenging scenario for social researchers to reach a conclusive

[2] See United Nations Official Home Page of SDG 16. https://www.un.org/sustainabledevelopment/peace-justice/

Table 3.3 Different targets and indicators under SDG-16

Targets	Indicators
16.1: Significantly reduce all forms of violence and related death rates everywhere	16.1.1: Number of victims of intentional homicide per 100,000 population, by sex and age 16.1.2: Conflict-related deaths per 100,000 population, by sex, age and cause 16.1.3: Proportion of population subjected to (*a*) physical violence, (*b*) psychological violence and/or (*c*) sexual violence in the previous 12 months 16.1.4: Proportion of population that feel safe walking alone around the area they live after dark
16.2: End abuse, exploitation, trafficking and all forms of violence against and torture of children	16.2.1: Proportion of children aged 1–17 years who experienced any physical punishment and/or psychological aggression by caregivers in the past month 16.2.2: Number of victims of human trafficking per 100,000 population, by sex, age and form of exploitation 16.2.3: Proportion of young women and men aged 18–29 years who experienced sexual violence by age 18
16.3: Promote the rule of law at the national and international levels and ensure equal access to justice for all	16.3.1: Proportion of victims of (a) physical, (b) psychological and/or (c) sexual violence in the previous 12 months who reported their victimization to competent authorities or other officially recognized conflict resolution mechanisms 16.3.2: Unsentenced detainees as a proportion of overall prison population 16.3.3: Proportion of the population who have experienced a dispute in the past two years and who accessed a formal or informal dispute resolution mechanism, by type of mechanism
16.4: By 2030, significantly reduce illicit financial and arms flows, strengthen the recovery and return of stolen assets and combat all forms of organized crime	16.4.1: Total value of inward and outward illicit financial flows (in current United States dollars) 16.4.2: Proportion of seized, found or surrendered arms whose illicit origin or context has been traced or established by a competent authority in line with international instruments

(continued)

Table 3.3 (continued)

Targets	Indicators
16.5: Substantially reduce corruption and bribery in all their forms	16.5.1: Proportion of persons who had at least one contact with a public official and who paid a bribe to a public official, or were asked for a bribe by those public officials, during the previous 12 months 16.5.2: Proportion of businesses that had at least one contact with a public official and that paid a bribe to a public official, or were asked for a bribe by those public officials during the previous 12 months
16.6: Develop effective, accountable and transparent institutions at all levels	16.6.1: Primary government expenditures as a proportion of original approved budget, by sector (or by budget codes or similar) 16.6.2: Proportion of population satisfied with their last experience of public services
16.7: Ensure responsive, inclusive, participatory and representative decision-making at all levels	16.7.1: Proportions of positions in national and local institutions, including (*a*) the legislatures; (*b*) the public service; and (*c*) the judiciary, compared to national distributions, by sex, age, persons with disabilities and population groups 16.7.2: Proportion of population who believe decision-making is inclusive and responsive, by sex, age, disability and population group
16.8: Broaden and strengthen the participation of developing countries in the institutions of global governance	16.8.1: Proportion of members and voting rights of developing countries in international organizations
16.9: By 2030, provide legal identity for all, including birth registration	16.9.1: Proportion of children under 5 years of age whose births have been registered with a civil authority, by age
16.10: Ensure public access to information and protect fundamental freedoms, in accordance with national legislation and international agreements	16.10.1: Number of verified cases of killing, kidnapping, enforced disappearance, arbitrary detention and torture of journalists, associated media personnel, trade unionists and human rights advocates in the previous 12 months 16.10.2: Number of countries that adopt and implement constitutional, statutory and/or policy guarantees for public access to information
16.a: Strengthen relevant national institutions, including through international cooperation, for building capacity at all levels, in particular in developing countries, to prevent violence and combat terrorism and crime	16.a.1: Existence of independent national human rights institutions in compliance with the Paris principles
16.b: Promote and enforce non-discriminatory laws and policies for sustainable development	16.b.1: Proportion of population reporting having personally felt discriminated against or harassed in the previous 12 months on the basis of a ground of discrimination prohibited under international human rights law

Source Collected from United Nations (2024)

remark on whether and to what extent the global power orders are in concordance with the peace process.

Next, recalling the SDG goals deadline, i.e., 2030, this is an anxious stage for SDG-16 if the world considers 'positive peace only' as 'peace' which arguably makes sense to be true. The world is witnessing a growing number of war victims, refugees, IDPs (internally displaced persons), stateless persons, and asylum seekers, including females and children. Fulfillment of positive peace, in this scenario, seems to be a faraway object. The first five matrices of the SDG-16 (i.e., 16.1 to 16.5) orbit around to enquire 'how many conflicts and violence are in place,' 'how many casualties are there' and 'how far corruption is the governance in dealing with public affairs.' All of those are yet far to be achieved for many parts of the world. World amidst the tremendous ascendency of neoliberalism, the competition between states to gain higher order in 'power' status, and the consolidation of major resources by a few, the remaining SDG matrices seem not the focus of many states to execute within the deadline.

Sustainable Development is, at its core, to ensure the benefits of development reach equitably across space (i.e., to the citizens residing in every corner of the world) and the justified actions that could guarantee the world resources are kept available across time (i.e., as it is now, so equitably it will be for future generations). There should not be any disagreements that prevailing peace (i.e., positive peace) for a prolonged time throughout the globe is a precondition to making this happen. While the march of history indicates a gradual imbalance of powers across time, in the present scenario of power order, there is no finite prediction of whether this peace will be possible by balancing powers. The true implementation of the 'culture of peace' is the demand of the time for which the United Nations should find alternative avenues. Peace for sustainability carries a much broader dimension than that of what is covered presently within SDG-16 matrices.

3.6 Nonviolence for Achieving Peace and Sustainability

> "If everyone fought for their own convictions there would be no war"
>
> (War and Peace, Leo Tolstoy, Vol 1, Part 2: 25)

The idea of achieving peace through practicing solidarity for which the Indian subcontinent has long been the beacon light to the world which is experiencing war, conflict, and violence in a frequent manner. The oldest channel of dissemination of the ideas of peace through solidarity was scripted in *Krishna Yajurveda Taittiriya Upanishad,*[3] *verse 2.2.2* which is estimated to have been composed around the sixth

---

[3] It is a primary *Upanishad*, a part of the *Yajurveda* which is a Sanskrit text composed in the Vedic era. The *Yajurveda* is basically large religious texts and is considered the oldest layer of Sanskrit literature. Interestingly, the *Taittiriya Upanishad* is one of the earliest known texts to have an index at the end of each section.

century BCE (Michael, 2007). A very popular *sloka* from this *verse* articulates the words of solidarity and decaying of animosity to bring peace as follows:

> Om [i.e., the Supreme God], May we be protected together,
>
> May we be nourished together,
>
> May we work vigorously together,
>
> May our intellect be sharpened,
>
> Let there be no animosity amongst us,
>
> Om! [Let instill the] peace [in me], peace [in nature], peace [in divine forces].
>
> *(Translated in English)*

By the next century, the origin of Buddhism in India and its expansion primarily following the trade route or, particularly through the ancient silk route and later overseas, emerged a refined idea of peace, especially in the East, as harmony, tranquility, serenity, quietude, and concord. Undoubtedly, Buddhism, among those of other world religions, on peace and violence accorded in creating new sets of values regarding the ways people manage conflict and maintain peace via nonviolent means through a holistic sense.

> The view of peace as a collective product is well in line with the Buddhist worldview based on the principle of dependent origination which emphasizes the mutual influence of all the elements involved in any situation. With this interdependent frame of reference, Buddhists would prefer a holistic view of peace, instead of peace in separate contexts such as schools, families, or the environment. This is again very close to what many peace studies scholars have advocated as the ultimate vision of peace.
>
> (Yeh, 2006: 93; see also Galtung, 1993; Galtung and Ikeda, 1995; Turpin and Kurtz, 1997; Brock-Utne, 1997)

The world again witnessed the efficiency of practicing non-violence during the early twentieth century through the most fundamental principle of the philosophy of Mahatma Gandhi (1869–1948) in the form of '*Ahimsa*.' Gandhi's philosophy of '*Ahimsa*' is the ideas of love, life, and creation as opposed to violence or '*Himsa*,' which causes hatred, death, and destruction. Gandhi believed that the universal human value of non-violence ought to be cultivated not merely at the personal level, but at the social, national, and international levels too to resolve conflicts—maybe personal, social, national, and international. Gandhi's idea of '*Satyagraha*' (i.e., the force of truth) is accorded with a moral equivalent to war and conflict (Das Gupta, 1968; Bose, 1981; Parekh, 1997; Bhaneja, 2007; Biswas, 2015). The humanistic grandeur of '*satyagraha*' as a method of resolving conflict and securing justice has been appreciated globally and it remains, arguably, the key factor of India's independence from British rule. However, Gandhi's philosophy is far more noble and vast than that of securing freedom for India. It is a timeless gift to the world.

> Nonviolence is itself normative. It applies to any situation. He [i.e., Gandhi] was a fearless advocate of the dignity of the human person. This involves recognizing the unique endowments of the human person-in-community and providing a foundation for Human Rights. He is therefore called the emancipator of the oppressed. He helped millions of the discriminated poor to discover meaning in life and live a life worthy of the true human calling. The clarity of the conviction of the transcendental goal of every man led him to affirm the dignity of the human person at every level.
> (Padhi, 2014: 28)

The world would probably need a prolonged time to achieve what Galtung (1986) states as 'positive peace' or the similar version by Jane Addams and Martin Luther King Jr. in their works, as well (Ramey 2021). Negative peace is simply the scenario of the absence of violence, maybe through a ceasefire, or an end to political or racial repression, or many other forms of physical or structural violence which is often regarded as 'true peace' to many. A peaceful (say calm) situation followed by the end of a war is what Galtung (1986) considers - to be only half of the goal because achieving 'negative peace' does not create structures that promote 'sustainable peace.' Positive peace prevails when the absence of violence in all its forms (negative peace) is endowed with a degree of social justice reached equally to all people. Positive peace may be established in many forms but its goal is to improve the lives of everyone. It thoroughly denies the structures and norms that lead to hate, violence, conflict, and war (Galtung, 1996; Ramey 2021). The ethos of 'positive peace' complies with the basic objectives of the Sustainable Development Goals (SDGs) which, in a nutshell, aims to transform the world in such a way that no one is left behind.

The route through which peace is achieved by upholding solidarity, cooperation, and collaborative actions while keeping the ideas of non-violence at the core, has prolonged history and many success stories as well. Initially, it emerged as a form of religious advisories for mankind to keep along the track of cooperation and concord to achieve peace by avoiding conflicts. The avoidance of animosity which has long been believed to be key to achieving peace, gradually shaped into more stronger form of Gandhi's '*Ahimsa,*' the power of which is witnessed and acknowledged by the world.[4] Ranging from the Buddhist principles of 'non-killing' to Gandhi's eternal philosophy of 'non-violence,' the efficiency of 'non-violence' as a instrument for sanity, rights, and justice gets wider acceptability to mankind. Non-violence is a war-risk-free alternative to achieving peace.

> Nonviolence is not only ethically superior to violence, it also can be stronger; but it is often diluted or contaminated by anger or enmity and its power undermined ... The word satyagraha is often most loosely used and is made to cover veiled violence. But as the author of the word I may be allowed to say that it excludes every form of violence, direct or indirect, and whether in thought, word or deed. It is a breach of satyagraha to wish ill to an opponent or to say a harsh word to him or of him with the intention of doing harm. And often the evil thought or the evil word may, in terms of satyagraha, be more dangerous than actual violence used in the heat of the moment. Satyagraha is gentle, it never wounds. It must not be the result of anger or malice. It was conceived as a complete substitute for violence.

---

[4] A reading of 'The Power of Nonviolence,' by *Richard Gregg* is requested to get complete idea (See Gregg, 2018)

Gandhi (1933)

The softness of the idea is the true power of it. The idea is universal in nature which is equally true and justifiable in every corner of the world. So, it is not bounded by spatial and temporal frames. The path of *Ahimsa* that takes India to the path of **'Sovereign Socialist Secular Democratic Republic'**[5] is the same that helped Nelson Mandela in South Africa to obliterate apartheid[6] or the 14th Dalai Lama to speak for Tibetans and their culture.[7] While Mandela could bring change in South Africa, the Dalai Lama couldn't prevent Beijing from fighting. This is not to be treated as the 'win' or 'defeat' of the principle of *Ahimsa* or *Satyagraha*. Rather, while nonviolence is responded to violently, the question arises about the legitimacy of power safeguarding peace.

The path of nonviolence prescribes bringing the change first inside the mind, in the gesture, attitude, approach, and way of communication of an individual. This is a process like the purification of mind, will, and vision. A few sensitized minds could cater the positive thoughts to hundreds, pursuing the positive transformation of society, the states, and ultimately the world.

Nonetheless, nonviolence is itself an evolutionary process. It accords with the human history of civilization. The evolution of humans from the initial stage as hunter-gatherers to modern civilized social beings marks the path of the refinement of the values in human minds, sharpening of the level of intelligence to survive, and innovation of technologies to work efficiently for the well-being of mankind. The trajectory of civilization is marked by the transformations from animosity toward affection, from conflict toward cooperation, from isolation toward solidarity, and, from dissonance toward harmony. A transformation from violence toward nonviolence aligns along the same note. It is a natural action that fine-tunes with the direction of the evolution of human civilization from pre-historic times to the present. Persuasion of nonviolence is not merely fulfilling targets, rather it would be a march of humankind toward a sustainably peaceful world.

References

Bhaneja, B. (2007). Understanding Gandhi's ahimsa (non-violence). *Asteriskos, 3/4*(215–224), 216.

Biswas, P. (2015). Mahatma Gandhi's views on peace education. *Education Journal, Special Issue: Gender, Peace and Education, 4*(1–1), 10–12. https://doi.org/10.11648/j.edu.s.2015040101.13

Bose, A. (1981). A Gandhian perspective on peace. *Journal of Peace Research, XVIII*(2), 159–164.

Brock-Utne, B. (1997). Linking the micro and macro in peace and development studies. In J. Turpin & L. R. Kurtz (Eds.), *The web of violence: From interpersonal to global* (pp. 149–160). University of Illinois Press.

[5] See the preamble of the Constitution of India. Visit: https://legislative.gov.in/document-category/preamble-to-constitution-of-india/

[6] Please read 'Long Walk to Freedom: The Autobiography of Nelson Mandela' (See Mandela, 1994)

[7] A detailed study is available in 'Secret lives of the Dalai Lama: the untold story of the holy men who shaped Tibet, from pre-history to the present day,' by *Alexander Norman* (See Norman, 2010)

Das Gupta, A. K. (1968, December 7). Gandhi on social conflict. *Economic and Political Weekly, 3*(49).
Davies, S., Engström, G., Pettersson, T., & Öberg, M. (2024). Organized violence 1989-2023, and the prevalence of organized crime groups. *Journal of Peace Research, 61*(4).
Galtung, J. (1964). An Editorial. *Journal of Peace Research, 1*(1), 1–4.
Galtung, J. (1969). Violence, peace and peace research. *Journal of Peace Research, 6*(3), 167–191.
Galtung, J. (1985). Twenty-five years of peace research: Ten challenges and some responses. *Journal of Peace Research, 22*(2), 141–158.
Galtung, J. (1986). Peace theory: An introduction. Transcend International, 1986.
Galtung, J. (1993). *Buddhism: A quest for Unity and Peace.* Dae Won Sa Buddhist Temple.
Galtung, J. (1996). Peace by peaceful means: Peace and conflict development and civilization,.
Galtung, J., & Ikeda, D. (1995). *Choose peace: A dialogue between Johan Galtung and Daisaku Ikeda.* Pluto Press.
Gandhi, M. (1933). An impatient worker. *The Collected Works of Mahatma Gandhi, 60*, 381–382.
Global Peace Index. (2024). Institute of Economics and Peace. http://visionofhumanity.org/resources (accessed 10 Oct 2025).
Gregg, R. B. (2018). *The power of nonviolence.* Cambridge University Press. https://doi.org/10.1017/9781316671351
Grewal, B. S. (2003, August 30). Johan Galtung: Positive and negative peace. Building peace. *Forum.* https://www.buildingpeaceforum.com/no/fred/Positive_Negative_peace.pdf
Jervis, R. (1986). From balance to concert: A study of international security cooperation. In K. A. Oye (Ed.), *Cooperation under anarchy* (pp. 58–79). Princeton University Press.
Lasswell, H. D. (1965). *World politics and personal insecurity.* Free Press.
Mandela, Nelson (1994). Long walk to freedom: The autobiography of Nelson Mandela. Little Brown and Company,. ISBN 9780316545853.
Michael, M. S. (2007). The cyprus peace talks: A critical appraisal. *Journal of Peace Research, 44*(5), 587–604.
Norman, A. (2010). *Secret lives of the Dalai lama: The untold story of the holy men who shaped Tibet, from pre-history to the present day* (p. 9780385530705). Harmony.
Padhi, B. (2014). Meeting of minds: Gandhi and Mandela on non-violence. *Odisha Review, 2014*, 28. https://magazines.odisha.gov.in/Orissareview/2014/August/engpdf/27-34.pdf
Parekh, B. (1997). *Gandhi: A brief insight.* Sterling Publication Inc.
Ramey, B. (2021). *Positive peace.* Peacebuilding and Entropy. https://www.nonviolenceny.org/post/positve-peace-peacebuilding-entropy
Roche, D. (2003, April 02). The human right to peace by Senator Douglas Roche O C. In *Address to Liu Institute for global issues, hosted by the Simons Centre for Peace & disarmament studies.*
Roche, D. (2003a). *The human right to peace* (p. 9782895074090). Novalis.
Rummel, R. J. (1981). *Understanding conflict and war: Vol. 5: The just peace.* Sage Publications.
Schmid, H. (1968). Peace research and politics. *Journal of Peace Research, 5*(3), 217–232.
Schweller, R. (2016, May 09). The balance of power in world politics. Oxford research Encyclopedia of. *Politics.* Retrieved 8 Nov. 2024, from https://oxfordre.com/politics/view/10.1093/acrefore/9780190228637.001.0001/acrefore-9780190228637-e-119
Turpin, J., & Kurtz, L. R. (1997). *The web of violence: From interpersonal to global.* University of Illinois Press.
UCDP. (2024). Uppsala Conflict Data Program. https://ucdp.uu.se/
UNESCO. (1997, October 29). General Conference Twenty-ninth Session, Paris 1997: [Item 4.14 of the agenda]: REPORT BY THE DIRECTOR-GENERAL ON THE HUMAN RIGHT TO PEACE. https://unesdoc.unesco.org/ark:/48223/pf0000110027
United Nations. (1999, October 06). Resolution adopted by the General Assembly on 06 October 1999 [without reference to a Main Committee (A/53/L.79)] 53/243: Declaration and Programme of Action on a Culture of Peace. https://documents.un.org/doc/undoc/gen/n99/774/43/pdf/n9977443.pdf

United Nations. (2017, July 10). Resolution adopted by the General Assembly on 6 July 2017 [without reference to a Main Committee (A/71/L.75)] 71/313: Work of the Statistical Commission pertaining to the 2030 Agenda for Sustainable Development. https://ggim.un.org/documents/a_res_71_313.pdf

United Nations. (2000, September 18). Resolution adopted by the General Assembly [without reference to a Main Committee (A/55/L.2)]: 55/2. United Nations Millennium Declaration. https://documents.un.org/doc/undoc/gen/n00/559/51/pdf/n0055951.pdf

United Nations. (2002). UN Millennium Campaign. https://www.un.org/millenniumgoals/bkgd.shtml

United Nations. (2015b). Millennium Development Goals Report 2015. https://www.un.org/millenniumgoals/2015_MDG_Report/pdf/MDG%202015%20rev%20(July%201).pdf

United Nations. (2024). SDG indicators and Reports. https://unstats.un.org/sdgs

United Nations. (2015a, September 25). Resolution adopted by the General Assembly on 25 September 2015 [without reference to a Main Committee (A/70/L.1)]: 70/1 Transforming our world: the 2030 Agenda for Sustainable Development. https://documents.un.org/doc/undoc/gen/n15/291/89/pdf/n1529189.pdf

Wiberg, H. (1981). JPR 1964-1980: What have we learnt about peace. *Journal of Peace Research, 18*(2), 111–148.

Yeh, T. D.-l. (2006). The way to peace: A Buddhist perspective. *International Journal of Peace Studies, 11*(1), 93–94.

Mukunda Mishra is an associate professor in the Department of Geography and designated vice-principal of Dr. Meghnad Saha College in West Bengal, India. The college is affiliated with the University of Gour Banga. Dr. Mishra completed his postgraduate studies in geography and environmental management at Vidyasagar University and holds a Ph.D. in geography from the same university. He was selected for the National Merit Scholarship by the Ministry of Human Resource Development, Government of India. His research chiefly focuses on analyzing unequal human development and creating multi-criteria predictive models. He has more than a decade of hands-on experience in dealing with development issues at the ground level in various districts of eastern India. Dr. Mishra has, in his credit published one monograph and three edited research volumes so far from the house of Springer-Nature, and he has more than thirty-five research articles and book chapters published in journals and books of international repute.

Chapter 4
Environmental Politics in Ladakh, India: Voice and Counter-Voice

Somenath Halder

Abstract Metaphorically, the flag of 'growth, "profit-making, 'and 'mask-worn development' is hoisted and winded everywhere while notions *like* restoration or conservation are thrown back into the trash nowadays. The present chapter underlines the existing micro-environmental status of the Ladakh region (India) and its future prolific change, with the unwanted consequences of melting (surviving) glaciers, unprecedented ambient weather change (at the local level), and deterioration of ever-existing fragile ecology. This chapter mainly uses the data available in online sources from various news daily, journal articles, book chapters, blogs, and social media (in text, pictorial, and video format) platforms. After gathering the required information, the present section is proceeding with a logical way forward to reach the objective of synthesizing environmental politics in a sensitive hill environment of the Indian Himalayas. With the help of the voices of prime climate activists (Sonam Wangchuk), locals of the Ladakh region, and various experts, this chapter is attempting to highlight the issue and provide the best possible solutions.

Keywords Development politics · Climate change · Local environmental security · Geo-heritage restoration · Melting of glacier · Environmental extinction · Policy and reframed policy

4.1 Introduction

India, a country that boasts of being the largest democracy in the world, has been witnessing a rising concern of political disconnection prevailing between 'environmental issues' and 'election-winning agendas' (Chumikchan 2024; Times of India 2024). In India, electors are evidently not (still) sensitized to the burning topic of global climate change or the harsh calamities caused by unpredicted climatic events. Interestingly, an almost similar argument has been theorized in the global context that democratically elected personnel are less bothered for goal setting future outlook and

S. Halder (✉)
Department of Geography, Kaliachak College, Malda, West Bengal, India
e-mail: somnatgeo@gmail.com

M. Mishra et al. (eds.), *Contemporary Geopolitical Alignment in the Global South*,
https://doi.org/10.1007/978-3-031-81977-3_4

sensitization of environmental problems that confront the civilization (Heilbroner 1991; Ponting 1991; Zürn 1998; Baber 2004). However, amongst the published works, the enthusiastic focus is given to the theme of 'environmental politics' that is majorly imposed upon the world's perspective (Haynes 1999; Grove 2009; Rootes 2001; Kenis & Lievens 2014; Rydin 2014; Pellizzoni 2011, 2016; Blühdorn & Deflorian 2021; Marquardt & Lederer 2022). However, a few of them dealt with the pure ideological side of 'left' and 'right' wings on the subject area of 'environmental politics' in the name of 'green politics' and their parties' stand(s) (Carter 2013); politics on environment linked with Indigenous people (Dove 2006); inherited social movement from the grass-root level which may impact on global level environmental movement (Grove 1999) are important works encountered. Erstwhile, it is important to mention that beyond systematic and thematic literature review, the tune of the current chapter, and the previously conducted and established result of the similar environs, the following lines seemed interconnected. *Here it is*, far west (in a country *like* Chile) from the present concerned region, Ladakh, in the context of environmental politics, what Carruthers (2001,1) has quoted is still important for the present chapter.

> "…the lens of environmental politics reveals a disturbing underside to the Chilean miracle. Environmental policy, institutions, and participation are shaped and constrained by ominous legacies of history, dictatorship, and an economic orthodoxy inimical to sustainability. …. Chile's environmental movement likewise demonstrates promise and innovation but remains grounded in a civil society weakened and atomized by dictatorship and incomplete transition…"
>
> Carruthers (2001,1)

The above-stated comment is revealed as factual not only in the developing world (or the Global South) but also in the Western monopoly, or, say, the so-called 'developed world.' Before the in-depth assertion, we must define the terms 'environmental politics' and 'politics of the environment.' The very term 'environmental politics' states the study concerned with environmentally related political theories and ideas that evolved (with the passage of time), conceptualized the environmental stances of environmental movements and dominated political organizations, and, after all, met the impacts and results of implemented environmental policies in multiple-geopolitical entities (Ostrogorski 1902; Carter 2007, 2018; Dobson 2016; DeSombre 2020). However, another term, 'politics of the environment, 'differs slightly from the previous one. Ideologically, 'politics of environment' apprehend the progression of any political institutions (under any geographical entity) (Hays 1984) for solving any particular or holistic environmental problems with unwanted externalities connected to the over-usage of natural resources, which results in many ill-impacts on ecology and humans, instead of all suggested remedial measures (Silva 1998; Carter 2012; Switzer 2018). It means there are very few differences between the terms 'environmental politics' and 'politics of environment.'

Beyond the above-mentioned research attempts, very few studies are endorsed to discuss the sensitive mountain ecology, with a socio-political dimension, of India's extreme north. For example, Satterfield (2009), Goeury (2010), Sherratt (2013),

Kumar (2019), and TERI (2021). Meanwhile, it is important to note that none of the pre-mentioned research has touched on issues *like* environmental politics in Ladakh (India), a major research gap. Contrastingly, a large number of reports on established news daily have captured the limelight of Sonam Wanchuk's "climate fast, " which has been elaborately discussed in the subsiding sections.

The prime objective of this chapter is to focus on the *environmental politics* in the Ladakh region from a contextual perspective. Secondly, the current chapter is endorsed to find out the reality check Sonam Wanchuk's socio-political and cultural demand(s) after banning 370 (2019) in undivided Jammu & Kashmir (India). Thirdly, the chapter tries to suggest comprehensive (reframed) policy insights for the betterment of the diverse ecology of the Ladakh region and the citizens with varied religious backgrounds who have lived through generations.

The following subsection helps find the significance of the issue presently raised in the internationally and politically sensitive region under Indian Territory. Section 4.2 explains the types of data used and the methodological inputs to reach the mentioned objectives. Section 4.3 primarily focuses on the cause(s) for raising the topic (e.g., environmental politics) in an ecologically sensitive region *like* Leh, Ladakh (India). Under the same section, the subsiding sections seek to underline the details of the concerned study area and the reason behind the critical cross-discussion on environmental and political insurgence. The next section (Sect. 4.4) is intended to explain, in a broader sense, the overall discussion, while the connecting sub-sections are assigned to search for solutions for sustainable development goals in the Ladakh region and encounter the ongoing debates and recommendations from the author's point of view. Finally, in Section 4.5, the concluding remark has been made in favor of humanity.

4.1.1 Relevance of the Current Issue

Sometimes, it has been thought that what is going on in a country *like* India, in the name of economic growth, global economic ranking of the country, harnessing the resources, exploitation of natural and human resources, and many more, seems to be "all is well." But in the true sense of the term, what is going on may not be well, or in the long-term, it may be proved as a socio-political blunder. Generally, as common Indian stakeholders, we are not too bothered about the current transformation(s) beyond our local surroundings. Here is the catch. The political personnel knew the weak points of common Indian voters. In spite of solving the problems in cold desert environs (*like* the Leh-Ladakh region), for instance, promised statehood in the Ladakh region, and securities on various points *like* water, livelihoods, shelter, ecological, and religious conservations, the new reframed policies may be focusing national interest rather than regional interest. In connection with the present *chapter*, the pin-point focus is enlightened to discuss the complex and conflicting scenario of achieving 'sustainable development goals' (SDGs), the nation's priority on the global fiscal 'rat-race,' and prioritizing the demands of Ladakh's people. The above-stated discussions, in wholesome, are thought to be enough to underline the relevance of the current agenda.

4.2 Methods and Data

Unmatched with the contemporarily statistically model-based research and hybrid kind of AI-generated research, the present *chapter* has purposively adopted the classical methodological input to reach its pre-stated objectives. The data used in this chapter is primarily second-hand data and, in some cases, a tertiary database and or metadata. In an elaborative format, the concept of 'environmental politics,' or, say, 'politics of environments,' is established using already published research articles, books, and book chapters. Secondly, to record the existing ecological setup of the Ladakh region (India), various diversified data sources, from research articles to administrative reports, have been utilized. Thirdly, for critical re-explanations of the current persuasion, a cross-examination method has been adopted with the available news-daily reports, published articles, and the views of political personnel, as well as the ground-level climate victims, have been taken into account. After amalgamating all these ensemble data systematically, the current *chapter* is drafted with a futuristic view of the holistic welfare of Ladakh's environment and its inhabitants. Simply, Fig. 4.1 depicts the methodological structure of this chapter. On the other hand, ArcGIS software, with the help of the DEM model, was used to build the study area map.

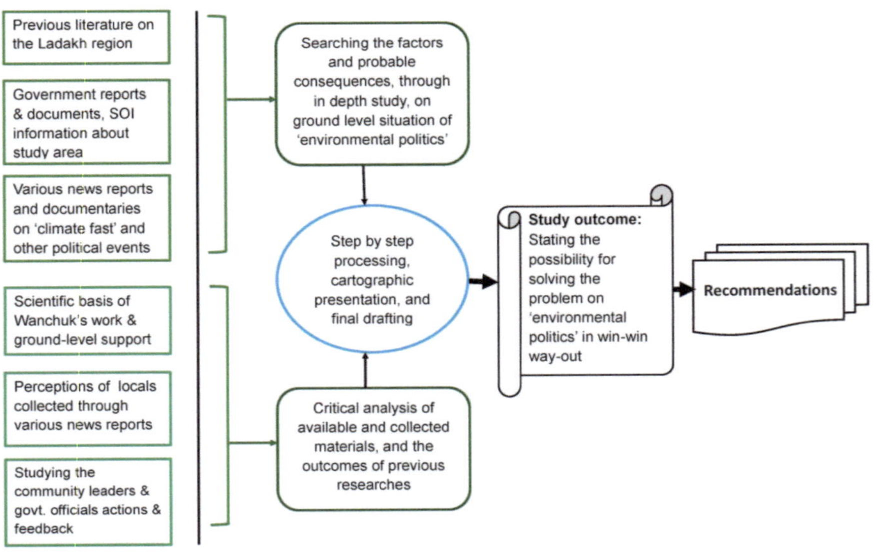

Fig. 4.1 Methodological frame of the study. *Source* Developed by author, *after* Kumar, 2019

4.3 Reason behind Current Talk on Environmental Politics

Before banning Article 370 (2019) in undivided Jammu & Kashmir (Indian part), it was a well-known fact that the Leh Ladakh provinces are the ever-neglected geophysical corner (BBC 2019; The Economic Times 2023). Neglect might be considered in the sense of any kind of infrastructural development and overall upliftment of the quality of life of Ladakh's inhabitants. And the viable reasons would be harsh climatic conditions (*like* one of the cold deserts of India), less concentration of voters, the dilemmas in decision-making for marginal corners, and many more. Abruptly after the demolition of the special status (which is temporary) of Jammu & Kashmir, Ladakh came into the limelight of Indian internal democracy. However, there was an unofficial or informal talk or mutual understanding and consideration (rather, it should be called hope) among the locals that in the near future, their fate was going to be transformed in a good direction. But, alas, all are gone in vain. In the face of a re-boosted developmental approach in Ladakh's environment, nearly everything is going to be altered, which may harm the local ecological setup (on a small scale) and in broader aspects of the earth's ecological diversity, as well as the current climate change phenomena. Here lies the significance of the current talk, which is clarified in the subsequent paragraphs.

This major section (Sect. 4.3) has two subsections, which contributes towards the understanding of the concerned area and its environmental legacies, and the contemporary politics of environment in the Ladakh part, which is located in the extreme north of Indian continent.

4.3.1 Ladakh and its Environs

Ladakh is constituted in the mid or central part of the former Jammu & Kashmir (India). Geographically, the concerned environmentally sensitive (micro) region is positioned in the crowned part of India. In the case of round figure assessment, this very region extended between 32°–36° North (latitude) and 75°–80° East (longitude) [specifically, the location of Leh is34° 12′ 34.2540′‘N and 77° 36’ 54.4032″ E.]. Figure 4.2 shows the overall environment of the study area. The geographical details of the study area can be seen in Fig. 4.3. Well recognized as cold desert environs, Ladakh, and uppermost flat topography having an altitude of 3000 m. The total area coverage is 45,110 sq. km. This concerned region (Ladakh) is separated into Kargil and Leh districts. Kargil's first district stayed at a height of 2750 m and Leh at 3505 m (Dame 2018). Experiencing massive temperature fluctuations (throughout the year) caused by geographical location, altitude, and rain-shadow pocket under western Himalaya, while in the discussed region, diurnal temperature ranges between −30 ° C to above 0 ° C (in winter season, the lowest temperature recorded ever as low as −40 ° C), and annual average precipitation is10cm (Demenge 2007; Ahmed et al. 2010; Kumar 2019; Sharma et al. 2022). The major geophysical features are winter with prolonged nature, rugged mountain terrain, water scarcity from natural sources, rainfall deficit, cultivable fertile land scarcity, and so forth. In the case of flora and fauna, as this region tolerated the cold desert condition, sparse vegetation ringed supreme, while trees *like* apples, walnuts, and apricots bloomed in the summer season. On the other hand, the common animal species are robins, snow leopards, red stars, Tibetan snow-cock, ravens, and hoopoes (Satterfield 2009). Among the valuable mineral resources, borax (available in Puga Valley), prime quality marble (available in Leh and Kargil), limestone, and granite are important. Moreover, other mineral resources (belonging to the Archean to Post-Cretaceous era) are yet to be extracted (Agarwal et al. 2021).

This union territory, constitutionally declared after banning Article 370 (2019), has some unmatched socio-demographic, economic, and socio-political features, described as follows. Here, 1.7% (274,289) of the Indian population resided within the total population of the country (according to Census 2011), among which major proportions are Muslim and Buddhist. The dominant languages locals speak are Ladakhis, Purit, Balti, English, and Tibetan (Agarwal et al. 2021; TERI 2021). Cultivation is the major economic activity practiced by Ladakh's people, where the region is constrained to fertile land and suitable climatic conditions (Humbert-Droz & Dawa 2004). Tourism is the next vital economic activity, followed by mineral resource extraction and other allied economic activities (TERI 2021). After the reformation of Ladakh as a union territory (under former Jammu & Kashmir), the socio-political governing authority headed on the secretariat level, involving local level of

Fig. 4.2 Panoramic view of Leh, the capital of Ladakh (*Courtesy*: Munkunda Mishra; captured on 21st April 2024)

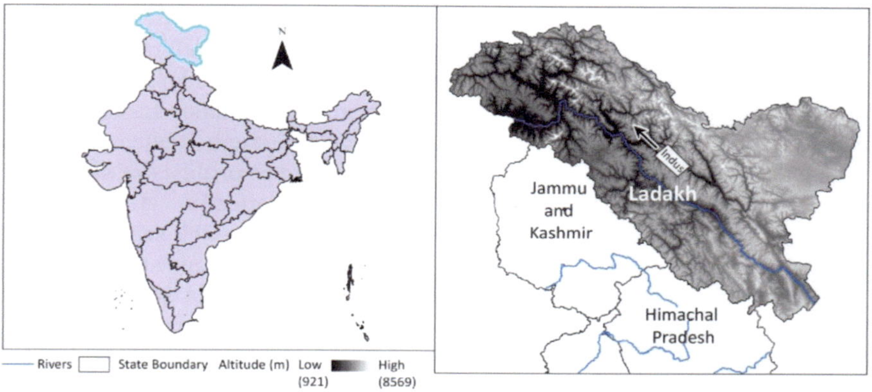

Fig. 4.3 Study area, Ladakh and its surroundings (N.B. Area coverage of Ladakh region is instructed mandatorily to show the entire area by SOI)

community leaders or, say, representative(s) and NGOs. And executively, here lies the problem concerned with local-level environmental politics. Detailed explanations are given in the subsequent sections.

4.3.2 *Emergence of Environmental Politics in Developmental Prosecution*

What is happening in present times, in the past, or the future in such a politically-attached country *like* India remains untouched from (any form of) political influences seems half-truth. Before the parliamentary decision to withdraw (temporary) special status of former Jammu & Kashmir, the life of the people of Leh and Ladakh region carried its own way with slow-motioned developments, *likewise* surviving with limited resources, lack of infrastructural progression, water scarcity, poor electricity supply, dependency on minimum agricultural production (especially centered in valley floor), and conventional mode of tourism. Immediately after the banning of Article 370 (2019), the locals of Ladakh start dreaming of "acche din" (translation: good times). But now, it has been revealed that their utopian dream(s) has/have been broken. Considerably, it should be well-accepted that neither any form of good nor positive change(s) can be possible overnight. Here's the reason a prominent figure of the concerned region, Sonam Wangchuk, who is not only an environmental activist but also a ground-level reformer who consistently works hard for the betterment of Ladakh's people and environment, took into consideration for worthy analysis. According to some important sources *like* YouTube channels (Wangchuk 2015, 2021a, 2021b), among several developmental activities and Indigenous reformations, he has successfully performed tremendous ground-breaking problem-solving tasks, such as ‘Ice Stupas’ (for mitigation of water crisis), Ladakh’s weather-friendly ‘solar-heated-tents’ for Indian Army and tourists. Besides these, he hasreformed numerous ground-level *works like* play-and-learning science for Ladakh’s children, escalatinge-ducational and environmental (restoration) awareness for common Ladakh people, and many more (Ahuja 2024; Gupta 2024). It would be the first attempt where a rational analysis is orbiting around one or two eminent personalities of the Ladakh region (India). On March 06, 2024, Sonam Wangchuk started his ‘climate fast’ of 21 days, followed by ten days fast by Ladakh’s women for the statehood, safeguarding delicate Himalayan ecology and overall development, and also to draw the attention of the central government of India (BBC 2024; Dolma 2024; ET Online 2024; Jain 2024; Mingle 2024). All of a sudden, *he* was also capable of attracting the limelight of well-established printed and motion media houses from local to global levels (Bhattacharya 2024). Modern media houses (e.g., BBC, The Times of India, The Economic Times, India Today, The Wire, and Digpu News Desk) began to take coverage of the peaceful socio-political and environmental movement against the newly reformed governance. Alongside this, a huge number of locals and environmental and climate activists (in support of Sonam Wangchuk) from the Ladakh region joined the ‘climate fast.’ Spontaneously, it has become a bigger topic (and also a debatable one) concerning environmental politics under the political geography of India, and further, it has reached the global platform as the center of coffee table discussion. Thoughtfully, it can be said that the reason behind the recognition and support (at national and international levels) is due to its concrete and grassroots protestand the long-term (previous) hard work of the *said* personality.

The basic points, in short formation, of demands revealed from the above-mentioned‘climate fast’ (in the shape of a peaceful hunger strike based on a Gandhian ideology of *Satyagraha* before India’s independence) are described as follows. (1) Protection of land rights (safety from outsider, industrial, and military encroachment against geoheritage safety and security of local lifestyle), (2) local level job reservation (ensure employment of local youths in government and administrative sectors), (3) selection of representative from Ladakh’s grass-root (despite outsider representative belonged to distant place unknown to problems and prospects of the very region), (4) preservation of Ladakh’s culture (restriction of intrusion of Cosmo culture which may altered the indigenous cultural dignity of the people of Ladakh), (5) conservation of regional environment (ensures the conservation of overall fragile ecology of Ladakh), and (6) demands of Sixth Scheduleinclusion provisioned under Indian Constitution (Mallick 2025).

Another eminent figure, Godbole (2024), advocating the current initiative made bySonam Wangchuk (Magsaysay award winner) of the Ladakh region, went on a three-week hunger strike and demanded regional autonomy. Meanwhile, heemphasized issues *like* the melting of the glaciers, threatened Indigenous tribal culture, (unexpected) negotiation with currently chaired governance, and many more. According to Ghai (2024), Sonam Wangchuk re-coined the aftermath of the adoption of popularized strategies for further development and so-called growth, remembering the long-term effect. For this purpose, he proposes (according to the author’s opinion) ‘stop and go determinism,’ also called ‘neo-determinism.’ Side by side,

Singh (2023) addresses the issue of climate-related vulnerability in the Ladakh region, especially highlighting the flash flood and cloud burst scenario due to the current consequence of 'global climate change'(GCC). Chandra's (2019) viewpoint states that if the concerned region is declared a union territory (UT), it will benefit from the conservational viewpoint regarding ecological balancing and socio-cultural sustenance. While in reality, it was not.

Instead of all these, an additional voice raged against the central government, saying that the administration did not keep its word on autonomy statehood and other relevant agendas. The face of the protest is Sajjad Kargili, who invites the Buddhists and Muslim groups belonging to the Ladakh region to join the mission of snatching their demand (Sharma 2024). For a more specific sum-up remark, it would be worthwhile to mention the statement(s) made by Sonam Wangchuk regarding the current environmental politics during the peaceful hunger strike (Mingle 2024).

> *Statement 1*: "Over my lifetime, I have seen glaciers vanish. Glaciers that used to be right next to roads have retreated hundreds of meters."
>
> *Statement 2*: "My appeal always to people out there is, please live simply in your big cities, so we in the mountains may simply live."

In short, it can be stated that the current emergence of politics on the environment in India, particularly in the Ladakh region, underlines the critical interplay between aspiration aroused from the local level and the imposed national policies (keeping in view of national interest and monopoly governance). Contrarily, the peaceful and democratic way of protest and rally led by Sonam Wangchuk and other local community leaders, solely supported by local people, especially women, accentuate the rising unhappiness among local dwellers regarding unfulfilled commitments on the issue of statehood (grass-root level autonomy) and justifiableimprovement post the revocation of Article 370 (2019). The above-discussed socio-political and environmental movement is able to catch the attention of both nationwide and universal scales, which replicates a deeper tautness between so-called 'modern development' and the conservancy of Ladakh's diversified culture and heritage.

4.4 Discussion

Before starting a discussion on the corresponding topic, it is viable to coin the statement made by Sonam Wangchuk to a Deccan Herald news reporter.

> "Unless we as a nation make our wishes clear, politicians will always be swayed by immediate, instant gains over long-term losses, and long-term gains if they do things well" (Pailoor 2024: 1)

It means it is high time, as common citizens, that we (the voters) be aware of the hidden intentions of anonymous politicians behind any promises or gimmick actions they are trying to make. Side by side, the common voters also should deeply understand the intention of any announced policy and its after-effects in the short and long term. In support of the above statement, Bhat (2024) has re-quoted the speech of Sonam Wangchuk that the bureaucratic system is presently unwilling to know how or that it is beyond their capabilities to see the complexities continue in Ladakh's environs.

In the core part of the discussion, it can be said that, in the said corresponding studied region, a symbolic and recognizable fight (against the chaired governance) for acknowledgment, autonomy in the form of statehood, and the fundamental rights (given by the Indian constitution) has unfolded. The said movement was led by the eminent personality of the very region, who undertook 21 days of climate starvation in tough wintertime conditions. Theoretically, Wanchuk's "climate fast" is truly motivated by Gandhi's principles of nonviolent confrontation, capable of attracting attention to Ladakh's want for couturier governance, principally after the 2019 cancellation of its special status under Article 370. The tactical standing of the concerned region, especially from the perspective of the India-China border psychological and ground-based territorial strains, has commanded the amplification of military occurrence, which indirectly conveysconservationalchallenges attached to sensitive ecology and elegant hill culture. On the same side, the common dwellers of Ladakh (inhabiting this region of the heart of Himalaya through generations) are supporting the demands of insertion in the Sixth Schedule of the Constitution of India or in the form ofpermitting of statehood to guard their inherent rights and guarantee that development compliments their local urges. This nonviolent socio-political and environmental protestechoes a largermelee for balancing countrywide interests with the protection of local cultures and environments (Ayaan 2024; Lhaskyabs 2024). In accordance with the above-said analysis, it is commendable that the demands of local inhabitants are significant, and the peaceful protest is in favor of their inhabiting environment and the future restoration of inculcated culture and heritage. In due course, the Chairman of the District Advisory Committee on Environment (DACE), after a successful meeting (with the stakeholders of Ladakh), has declared the committee is harnessing the solution for conserving the regional environment. The said committee has taken some wise decisions to ensure the conservation of precious water (in the cold desert), encouraging artificial glaciers, reformatting and maintenance of local dry toilets (in households and public areas), outsourcing for financial crisis regarding the well-function ability of toilets, cleanliness and hygiene awareness program, and sustainable ecotourism (Department of Information & Public Relations, Leh, Ladakh [February 19], 2020). This indirectly hints that the local authority is already awakened and concerned about the ground-based problem a little bit. Oppositely, Bhardwaj (2024) made an argument from a different angle. He encounters during the ongoing protest in Ladakh focusing (majorly) on protection under the Sixth Schedule, some government officials opined the very schedule would provide grander autonomy (or, say, statehood) in principle, but its own has severe shortcomings.

Beyond this, after a rigorous review and deep assessment, it has been revealed that the demand for statehood of union territory, *likewise* Ladakh, is clashing with the central government standpoints. The points of barriers to giving statehood to the Ladakh region are as follows: (i) security concerns at the national level (as the region is surrounded by many disputed borders of neighboring states *like* China,

Pakistan, and Afghanistan), (ii) sovereignty and the idealistic unity at the national level, (iii) agenda of contemporary development (aspirational gaps between regional and national levels), (iv) constraints at the constitutional and legal ground (a positive desire of change may adhered a complex legislative process and also may face political resistance), (v) point of good governance and efficient administration (on which the very region is lacking off), and lastly (vi)environmental concerns (Ayaan 2024). Above all, searching for eco-friendly solutions and some negotiable suggestions are given in the following sections.

4.4.1 Finding the Modal Way out for SDGs in Ladakh

The major issues faced by the people of Ladakh in the present day's situation amidst the political and environmental doldrums are (a) restoring fragile and delicate ecology of Ladakh, (b) new challenges and coping strategies facing 'Climate Change Phenomena' (CCP), (c) way of suitable infrastructural development, (d) non-decentralized governance, (e) transformation of domicile policy, and (e) shortage in financing or opportunity in capital investment. The aforesaid findings are the outcome of this current *chapter*, derived after a rigorous literature review of the available literature, various established news-daily reports, feedback, and comments recorded from political personnel, bureaucrats, and the original landsmen from the Ladakh region. In Fig. 4.4, one can easily see the holistic (modified) model for sustainable development goals (SDGs) in the Ladakh region—situated at the heart of the Himalayas.

4.4.2 Debate and Recommendations

From a locally circulated news desk (in the Himalayan part), it has been evidenced that a new kind of controversy and conflict(s) have aroused regarding the policy of industrial land allotment, which may (further) challenge regional autonomy and the especially focused environmental concerns (Voice of Ladakh Desk 2023). As in the earlier section, it has been mentioned that due to the industrial land allocation and military land acquisition, there is a real possibility of alteration and erosion of Ladakh's culture and further fragmentation of the land of hill people. Secondly, due to the lack of fertile cultivable landand scarcity of irrigation facilities in cold desert regions, 'tourism' is the only means of economic sustainability. Instead, the growing tourist influx further enhances the region's water crisis. Thirdly, the demand for regional autonomy (statehood) for suitable decision and policy making (*likewise* Northeastern provinces of India), representing and prioritizing ground-level real problems, should be enacted. Meanwhile, in the case of the third issue, according to Tiwary (2024), Home Minister Amit Shah made it clear that it would not be possible for the government to accommodate the region's demand to be included in the Sixth Schedule of the Indian Constitution. Fourthly, in the near future, the incursion of

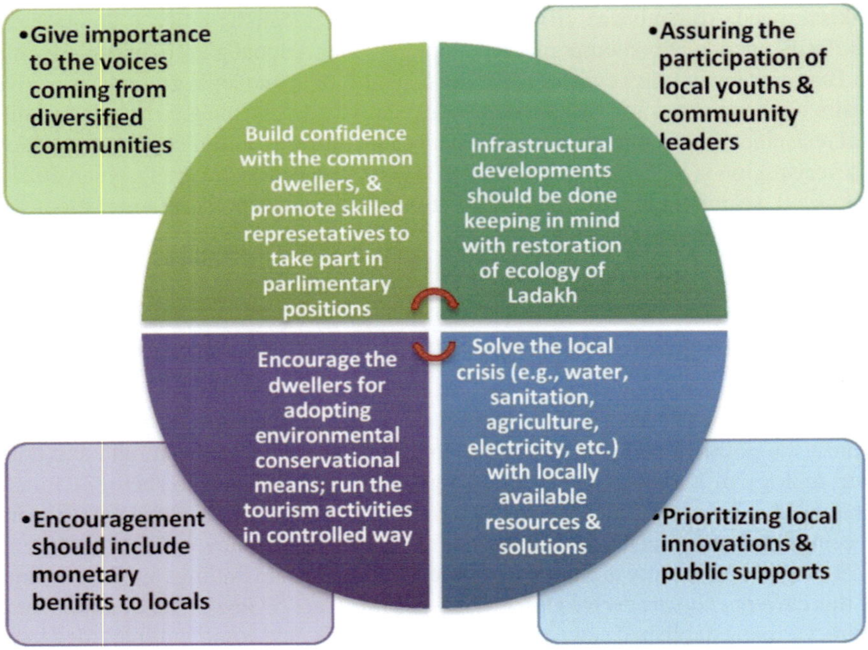

Fig. 4.4 A model prescribed (and modified by the author) SDGs in the Ladakh region

job-seekers from the rest of the part of India may harm the local job opportunities (which are already in crisis) and enhance frustration among Ladakh youths. Paljor (2024) already mentioned that the frustration among Ladakh's youths encourages them to take part in the "climate fast" event led by Sonam Wangchuk.

In a simplified or generalized manner, the recommendations are given as follows.

- Any policy implementation should be based on the confidence of the local dweller-sand community leaders, and it should also consider upcoming hurdles stated by stakeholders of the region.
- Make the provision for attaching the youths of Ladakh to any developmental projects *initiated by the Borde*r Road Organization (India), such as new road construction or expansion of existing roadways.
- In monitoring ongoing projects, we should employ Ladakh's youth on a pay-roll basis, which will satisfy the employment (partially) and prove the angle of sensitivity of the administration.
- It should give importance to the already successful solvation of problems *like* water crises (such as 'Ice Stupas')and climate-friendly shelters for commoners and army men (such as 'solar-heated tents') and also encourage those projects and campaigns for the participation of Ladakh's youth.

- Restrictions should be imposed on (uncontrolled) tourism activities, which have led to more water crises in the said region. The restriction should be imposed on tourists' permits, especially in the dry season.
- The next level of harnessing natural resources for driving the country's 'growth engine'should be done with the least possible deterioration of Ladakh's elegant ecology.
- As an innovative coping strategy for climate change phenomena and water crisis management, 'dry toilet' usage and environment-friendly waste management should be encouraged in a broader sense. In the coming days, the Ladakh region will be the *torch bearer* in the (said) subject throughout the country.
- Several welfare schemes run in the rest of the country should also be effectively implemented in the said region without any document-oriented complexities.

4.5 Conclusion

India's crown part, located north of the Himalayas, is going through a ground-attached tussle (in the shape of nonviolent protest) with the centrally chaired administration. This peaceful movement is fought for the environment in the delicate Himalayan region and also for the people living therein. The key points of the socio-political and environmental movement are regional recognition, land rights and decision rights and liberty, preservation of cultural heritage and geoheritage, ease accession of beneficiary schemes, and so forth. The ever-discussedpeaceful movement, named "Climate Fast," was commanded by Sonam Wangchuk and reinforced by the joint efforts of communities that inhabited the Ladakh region. Meanwhile, this strong effort initiated by the said person automatically arrests the attention of the country's media and global media. Their demands concern activities in running mode compacted with environmental protection and climate change centric. The second major demand attached to the boosted infrastructural development must include ecological sensitivity and local-level aspiration. The third and prime demand underlines the regional autonomy for any environment-friendly, community-supporting (reframed) policy initiatives.

As per the current author's neutral viewpoint, the region (Ladakh) is sensitive to ecological, socio-cultural, and anthropological dimensions. Presently, dimensions *like* geopolitics add a new, worried-some aspect to the already complex interplay between the original landsmen and the governing system. In a more highlighted format, it can be said that the *said region* is already situated on the verge of countries eagerly waiting to encroach or invade by any means. Thus, with respect to the situation mentioned above, the Indian government should handle the problems of the newly formed union territory and address the aspirations of the local people of Ladakh. Suppose any kind of dilemma or contradiction (in the form of misunderstanding) happens between the government, bureaucrats, community leaders, and local people. In that case, it will demolish interconnected threads of ecological conservation, national sovereignty, and sustainable development.

Acknowledgement First of all, I would like convey my gratitude to the editors for giving such an opportunity for describing my viewpoints on the concerned topic. Secondly, I would like to greet those anonymous reviewers for their constructive comments and suggestions. Thirdly, I also indebted to my scholar Rajesh Sarda for his entire support, and Dr. M. Mishra for providing the image of Leh, Ladakh. It would be unethical if I could not acknowledge the contribution of previous researchers worked on this field.

References

Agarwal V, Sood D, Mittal AS, Gyaltson T, Varma N, & Jain S. (2021, September).: Unleashing potential. ASSOCHAM. https://surayaan.blogspot.com/

Ahmed, N., Higgins, A., & Norphel, C. (2010). *Snow water harvesting in the cold desert Ladakh: An introduction to the artificial glacier project*. Leh Nutrition Project.

Ahuja S. (2024, June 17). The climate of craft in Ladakh. The voice of fashion. Retrieved from: https://www.thevoiceoffashion.com/sustainability/features/The-Climate-of-Craft-in-Ladakh-5968 Accessed 01 Jul 2024.

Baber, W. F. (2004). Ecology and democratic governance: Toward a deliberative model of environmental politics. *The Social Science Journal, 41*, 331–346. https://doi.org/10.1016/j.soscij.2004.04.014

BBC. (2024, March 22). Ladakh: The thousands of Indians protesting in freezing cold. BBC News. Retrieved from: https://www.bbc.com/news/world-asia-india-68573281. Accessed 03 Jul 2024.

BBC. (2019, August 6). Article 370: What happened with Kashmir and why it matters. Retrieved from: https://www.bbc.com/news/world-asia-india-49234708 Accessed 03 Jul 2024.

Bhardwaj A. (2024, April 20). Why Ladakh is angry with central govt — Power dynamics, joblessness & ecological concern. The print. Retrieved from: https://theprint.in/india/why-ladakh-is-angry-with-central-govt-power-dynamics-joblessness-ecological-concern/2046546/. Accessed 04 Jul 2024.

Bhat G. (2024, March 25). New Delhi's bureaucracy can't understand complexities unique to Ladakh: Sonam Wangchuk. The Hindu (Business Line.).Retrieved from: https://www.thehindubusinessline.com/news/new-delhi-based-bureaucracy-cant-understand-complexities-unique-to-ladakh-sonam-wangchuk/article67990575.ece. Accessed 02 Jul 2024.

Bhattacharya S. (2024, April 4). Icy desert Ladakh turns up the heat on Modi ahead of Indian elections. The diplomat. Retrieved from: https://thediplomat.com/2024/04/icy-desert-ladakh-turns-up-the-heat-on-modi-ahead-of-indian-elections/ Accessed 01 Jul 2024.

Blühdorn, I., & Deflorian, M. (2021). Politicisation beyond post-politics: New social activism and the reconfiguration of political discourse. *Social Movement Studies, 20*(3), 259–275.

Carruthers, D. (2001). Environmental politics in Chile: Legacies of dictatorship and democracy. *Third World Quarterly, 22*(3), 343–358. https://doi.org/10.1080/01436590120061642

Carter, N. (2013). Greening the mainstream: Party politics and the environment. *Environmental Politics, 22*(1), 73–94. https://doi.org/10.1080/09644016.2013.755391

Carter N (2nd edition.). (2012). The politics of the environment. Cambridge University Press. doi:https://doi.org/10.1017/CBO9780511819179.

Carter, N. (2007). *The politics of the environment: Ideas, activism, policy* (2nd ed.). Cambridge University Press.

Carter, N. (2018). *The politics of the environment: Ideas, activism, policy*. Cambridge University Press.

Census. (2011). *Census of India: Basic population figure of India and states, 2011*. Bureau of Census Operation and Statistics. https://censusindia.gov.in/census.website/data/population-finder

Chandra R. (2019, August 07). View: Ladakh will greatly benefit from government's decision to make it into a union territory. The Economic Times. Retrieved from: https://economictimes.indiatimes.com/news/politics-and-nation/view-ladakh-will-greatly-benefit-from-governments-decision-to-make-it-into-a-union-territory/articleshow/70577272.cms?from=mdr. Accessed 04 Jul 2024.

Chumikchan R. A. (2024, April 24). Environment in elections: Will climate change alter voting dynamics in Ladakh? Down to Earth (Climate Issue). Retrieved from: https://www.downtoearth.org.in/climate-change/environment-in-elections-will-climate-change-alter-voting-dynamics-in-ladakh%2D%2D95761 Accessed 10 Jul 2024.

Dame, J. (2018). Food security and translocal livelihoods in high mountains: Evidence from Ladakh, India. *Mountain Research and Development, 38*(4), 310–332.

DeSombre. (2020). *What is environmental politics*. John Wiley & Sons.

Demenge, J. (2007). Measuring ecological footprints of subsistence farmers in Ladakh. In International Ecological Footprint Conference, Cardiff, Wales (UK). http://web.mnstate.edu/robertsb/307/ANTH (Vol. 20307).

Dobson, A. (2016). *Environmental politics: A very short introduction*. Oxford University Press.

Dolma T. (2024, April 9). Ladakh's urgent cry for environmental protection. The Sunday magazine. Retrieved from: https://www.dailypioneer.com/2024/columnists/ladakh%2D%2D-s-urgent-cry-for-environmental-protection.html. Accessed 12 Jul 2024.

Dove, M. R. (2006). Indigenous people and environmental politics. *The Annual Review of Anthropology, 35*(1), 191–208.

ET Online. (2024, March 26). Sonam Wangchuk ends 21-day hunger strike, demanding statehood for Ladakh, The Economic Times. Retrieved from: https://economictimes.indiatimes.com/news/politics-and-nation/sonam-wangchuk-ends-21-day-hunger-strike-demanding-statehood-for-ladakh/videoshow/108798246.cms?from=mdr. Accessed 12 Jul 2024.

Ghai R. (2024, March 24). Ladakh wants to prevent what happened in Himachal Pradesh, Uttrakhand & Sikkim: Sonam Wangchuk. Down To Earth (Climate Change). Retrieved from: https://www.downtoearth.org.in/climate-change/ladakh-wants-to-prevent-what-happened-in-himachal-pradesh-uttarakhand-sikkim-sonam-wangchuk-95213.Accessed 12 Jul 2024.

Godbole T. (2024, April 19). India's Ladakh region demands autonomy. DW (Politics,). Retrieved from: https://www.dw.com/en/indias-ladakh-region-demands-autonomy-to-deal-with-climate-change/a-68873382. Accessed 7 Jul 2024.

Goeury, D. (2010). Ladakh, kingdom of sustainable development? Protecting the natural environment to protect identity. *Journal of Alpine Research, 98*(1) http://rga.revues.org/1147

Grove, R. C. (1999). Environmental movements: From local global. *Environmental politics, 8*(1), 1–12. https://doi.org/10.1080/09644019908414435

Grove, M. (2009). Hunter–gatherer movement patterns: Causes and constraints. *Journal of Anthropological Archaeology, 28*(2), 222–233.

Gupta K. (2024, June 13). The ice man of Ladakh building artificial glaciers in the Himalayas. BBC. Retrieved from: https://www.bbc.com/future/article/20240612-the-ice-man-of-ladakh-building-artificial-glaciers-in-the-himalayas. Accessed 06 Jul 2024.

Haynes, J. (1999). Power, politics and environmental movement in the third world. *Environmental Politics, 8*(1), 222–242. https://doi.org/10.1080/09644019908414445

Hays S. P. (1984). Three decades of environmental politics: the historical context. Retrieved from: https://www.osti.gov/biblio/5506746. Accessed 04 Jul 2024.

Heilbroner, R. (1991). *Inquiry into the human prospect*. W.W. Norton.

Humbert-Droz, B., & Dawa, S. (2004). *Biodiversity of Ladakh: Strategy and action plan*. ACB.

Jain A. (2024, March 21). Why Ladakh climate activist Sonam Wangchuk is on 'fast unto death' as climate protest enters 13th day. The Indian Express. Retrieved from: https://indianexpress.com/article/world/climate-change/why-ladakh-climate-activist-sonam-wangchuk-is-on-fast-unto-death-as-climate-protest-enters-13th-day-9220558/. Accessed 13 Jul 2024.

Kenis, A., & Lievens, M. (2014). Searching for 'the political' in environmental politics. *Environmental politics, 23*(4), 531–548.

Kumar, S. (2019). *A self-governance approach to solving the water crisis in Ladakh, India: The ice stupa project*. Master's thesis,. University of Twente.
Lhaskyabs S. (2024 March 06). Between protests and border incursions, insecurity is rising in Ladakh. The diplomat. Retrieved from: https://thediplomat.com/2024/03/between-protests-and-border-incursions-insecurity-is-rising-in-ladakh/. Accessed 15 Jul 2024.
Mallick, P. (Ed.). (2025). *Contours of democracy in new India*. Taylor & Francis.
Marquardt, J., & Lederer, M. (2022). Politicizing climate change in times of populism: An introduction. *Environmental Politics, 31*(5), 735–754. https://doi.org/10.1080/09644016.2022.2083478
Mingle J. (2024, March 28). At 11,500 feet, a 'climate fast' to save the melting Himalaya. Yale Environment 360. Retrieved from: https://e360.yale.edu/features/sonam-wangchuk-interview. Accessed 05 Jul 2024.
Ostrogorski, M. (1902). *Democracy and the organization of political parties*. Macmillan.
Pailoor A. (2024, April 3). India needs a green impact on electoral politics: Sonam Wangchuk, Deccan Herald. Retrieved from: https://www.deccanherald.com/india/ladakh/india-need-a-green-impact-on-electoral-politics-2960810. Accessed 09 Jul 2024.
Paljor J. (2024, May 23). In Ladakh, democracy and ecology take Centre stage. *Social Justice (IDR)*.. Retrieved from: https://idronline.org/article/social-justice/in-ladakh-democracy-and-ecology-take-centre-stage/. Accessed 09 Jul 2024.
Pellizzoni, L. (2011). The politics of facts: Local environmental conflicts and expertise. *Environmental Politics, 20*(6), 765–785. https://doi.org/10.1080/09644016.2011.617164
Pellizzoni, L. (2016). Catching up with things? Environmental sociology and the material turn in social theory. *Environmental Sociology, 2*(4), 312–321. https://doi.org/10.1080/23251042.2016.1190
Ponting, C. (1991). *A green history of the world*. Penguin.
Rootes, J. A. (2001). *The making of green knowledge: Environmental politics and cultural transformation*. Cambridge University Press.
Rydin, Y. (2014). The challenges of the "material turn" for planning studies. *Planning Theory & Practice, 15*(4), 590–595. https://doi.org/10.1080/14649357.2014.968007
Satterfield L. (2009). Trailing the Snow Leopard: Sustainable Wildlife Conservation in Ladakh (India). SIT Study Abroad, Spring 2009. (pp. 1–73).
Sharma B. (2024, March 28). As BJP Breaks promises of autonomy, Buddhists & Muslims join forces, protests rage in Ladakh. Article 14.Retrieved from: https://article-14.com/post/as-bjp-breaks-promises-of-autonomy-buddhists-muslims-join-forces-protests-rage-in-ladakh%2D%2D6605246391878. Accessed 09 Jul 2024.
Sharma, S., Tamchos, S., Kumar, R., & Sah, P. (2022). Low-cost mitigation strategies for community resilience in the trans-Himalaya to address food and water scarcity in changing environment. *Environmental Monitoring and Assessment, 194*(12), 879. https://doi.org/10.1007/s10661-022-10546-2
Sherratt, J. A. (2013). History-dependent patterns of whole ecosystems. *Ecological Complexity, 14*, 8–20.
Silva E. (1998). The politics of environment and development. Latin American Research Review 33(3): 230–247. JSTOR. http://www.jstor.org/stable/2504022. Accessed 01 Jul 2024.
Singh, R. (2023, September 18). *Ladakh's vulnerability to flash floods: Adaptation and mitigation measures*. Manohar Parrikar Institute for Defence Studies and Analyses.
Switzer, D. (2018). Politics and the environment. In A. Farazmand (Ed.), *Global Encyclopedia of public administration, public policy, and governance*. Springer. https://doi.org/10.1007/978-3-319-20928-9_2449
TERI. (2021). Carbon neutral and climate resilient Ladakh: A strategy document-action plan and roadmap. The energy and resources institute (TERI). (pp. 1–93). Retrieved from: https://www.niti.gov.in/sites/default/files/2023-03/Carbon%20Neutral%20Resource%20Efficient%20Strategy%20for%20Ladakh%20UT.pdf. Accessed 09 Jul 2024.

Times of India/Nagaland Page. (2024, March 17). Political disconnect: Why environmental issues not winning votes in India. Retrieved from: https://timesofindia.indiatimes.com/india/political-disconnect-why-environmental-issues-not-winning-votes-in-india/articleshow/108563256.cms. Accessed 01 Jul 2024.

The Economic Times. (2023, September 17). Ladakh saw unbelievable changes for the better after Article 370 abrogation: Lt Governor. Retrieved from: https://economictimes.indiatimes.com/news/india/ladakh-saw-unbelievable-changes-for-the-better-after-article-370-abrogation-lt-governor/articleshow/103732658.cms?from=mdr. Accessed 23 Jul 2024.

Tiwary D. (2024, March 17). Centre considers 'article 371-like' shield for Ladakh. The Indian Express. Retrieved from: https://indianexpress.com/article/india/govt-considers-art-371-like-shield-for-ladakh-9198036/. Accessed 25 Jul 2024.

Voice of Ladakh Desk. (2023, November 6). Controversy erupts over Ladakh's new industrial land allotment policy, challenging autonomy and environmental concerns, Voice of Ladakh. Retrieved from: https://www.voiceofladakh.in/2023/11/controversy-erupts-over-ladakhs-new-industrial-land-allotment-policy-challenging-autonomy-and-environmental-concerns/. Accessed 30 Jul 2024.

Wangchuk, S. (2015). *The monk, the engineer, the artificial glacier, ice stupa*. Sonam Wangchuk. www.icestupa.org

Wangchuk S. (2021a). The Ladakh tent, world's 1st solar heated military tent, made in India. Retrieved from: https://www.youtube.com/watch?v=IF-IYkU0gmQ. Accessed 28 Jul 2024.

Wangchuk S. (2021b). To make carbon neutral Ladakh project through sankalptaru website. Retrieved from: https://www.youtube.com/redirect?event=video_description&redir_token=QUFFLUhqbnFkLWlBX2w5cWUzREJNaDZzZy15b2ZnWEF6Z3xBQ3Jtc0ttT3dFUC1hR1haUklHdExHd0lVeWZKRjRtOTJkM09YaVhILXdmbTJ3amM3WWtaTUEzVkQtQ3lFMDlISDdHRE9NOHkzV1doUUFBTk51NDNlYXYzSS0wWS0yeEtUVEZyeFdadmNJV2tmWWZzWG1hbFZ3NA&q=%3A%2F%2Fsankalptaru.org%2Fcarbon-neutral-ladakh%2F&v=oBtSoWh1hJg. Accessed 28 Jul 2024.

Zürn, M. (1998). The rise of international environmental politics: A review of current research. *World Politics, 50*(4), 617–649.

Dr. Somenath Halder is an Assistant Professor in the Department of Geography at Kaliachak College, West Bengal, India. His academic focus is also come through sustainable development, environmental issues, climate impact assessmentand spatial analysis. He obtained his Ph.D. from the Visva-Bharati, Santiniketan, India. He has several publications in journals and books of international repute.

Chapter 5
China in Central Asia: Between Geopolitics and Cooperation

Jonathan Christian Dias dos Santos, André Santos da Rocha, and Pablo Ibañez

Abstract The twenty-first century has been characterized by the expansion of Chinese power and hegemony. Many of the strategies of this Asian country are centered on unorthodox and very varied foreign policy actions for each portion of the planet. The actions surrounding the Belt and Road Initiative, which is currently considered by experts as the largest axis of intersection of freight transport in the world, stand out. The Central Asian region is inevitably inserted on this route for its strategic position. This position gives the region not only China's maintenance and surveillance views, but also strategies to expand its influence through cooperation agreements.

Keyword Strategies · Belt and road initiative · China and Central Asia

5.1 Introduction

Since the late twentieth century, particularly following the end of the Soviet Union, the Chinese have been strengthening their diplomatic ties with Central Asian countries which were part of the Soviet bloc were Kazakhstan, Uzbekistan, Kyrgyzstan, Tajikistan, and Turkmenistan. The Sino-Central Asian relationship has been guided by cooperation agreements expanding in several areas such as the power sector, agriculture, military, culture, and the alike.

Cooperation has been the hallmark of Chinese diplomacy not only with Central Asian countries but rather with numerous states around the world. To the Chinese cooperation as a tool for diplomacy originates with the notion of a peace treaty offered after the end of the Cultural Revolution, and the economic and political reforms under Deng Xiaoping and spread throughout the twentieth and twenty-first centuries, which allowed China's entry in international organizations (Li & Worm,

J. C. D. dos Santos · A. S. da Rocha · P. Ibañez (✉)
Department of Geography, Federal Rural University of Rio de Janeiro (UFRRJ), Rio de Janeiro, Brazil
e-mail: ibanez.pablo@gmail.com

M. Mishra et al. (eds.), *Contemporary Geopolitical Alignment in the Global South*,
https://doi.org/10.1007/978-3-031-81977-3_5

2011). Through this political philosophy, Beijing has sewn cooperation agreements to increase its catalogue of economic and commercial partners.

In Central Asia there are two platforms for Chinese cooperation that were set up at different periods of China's contemporary economic rise. It was officially set up in 2001 and its first platform is called Shanghai Cooperation Organization; and the second one being the Belt and Road Initiative (BRI) in Kazakhstan announced in 2013. Such platforms feature functions and scales of action within the Chinese geopolitical strategy and they represent the expression of the Chinese geopolitical code which we shall explain its meaning further on.

China's geopolitical code is biased by the benevolence discourse of a community with one shared future (Ming, 2020), in which other nations should also have decision-making power on world order affairs. According to Western thinkers, the Chinese would be getting closer to the idea of "Soft Power" by Joseph Nye (Nye, 2012). In any event, the main feature of the Chinese geopolitical code—with the Shanghai Cooperation Organization and the Belt and Road Initiative—is the act of cooperation, that is to say, the act of assisting others and building relationships which, in theory, are mutually beneficial.

In this article, we seek to understand this contemporary Chinese geopolitical code, its way of acting—through cooperation agreements—and lastly, on how it is embodied in the territories of Central Asia. Therefore, we will divide this production into three sections. On the first section, we shall address the two main concepts of the text; the geopolitical code and cooperation—believed to be the main feature of China's geopolitical code.

In the second section, we will discuss the characteristics of the Shanghai Cooperation Organization and the Belt and Road Initiative platforms, where the Chinese geopolitical code is made operational; In the third section, we will demonstrate how the Chinese geopolitical code—the cooperation—is implemented in Central Asian territories through its main platforms of action.

The main objective of this text is to analyze the diplomatic relationship between China and Central Asia in the last centuries and how cooperation is the main feature of this relationship, and it also develops spatial imprints in Central Asian countries.

5.2 Theoretical Aspects Concerning the Chinese Geopolitical Code and Action in Central Asia

All world states have a relatively strong geopolitical relationship with each other. These relationships can occur at different scales and in different ways. Even if it is a close or distant relationship, all countries get engaged in different spheres and are constantly trying to attract or develop areas of geopolitical influence.

For some, the area of geopolitical influence is regional. For instance, Russia understands that the former Soviet Union space is an area in which it deems itself to have privileged interests (Medvedev, 2008). For others, the geopolitical sphere of influence is much wider, involving a global scale. Here, we can think of China and the United States, for example.

These geopolitical relations can also be understood as the geopolitical codes of countries. The geopolitical code, the main concept we use for this study, is mainly employed by scholars of Critical Geopolitics, a trend that has started in the context of the independence of the former European colonies in Africa in the 60s and 70s and gained even more prominence after the end of the Cold War for providing insight into territory and power relations, beyond the states-centric and unidimensional conception (Raffestin, 1993; Tuathail, 1999).

A code may be understood as "an elaborate model structure postulated as the rule underlying a series of concrete and individual messages which are suitable for it and only toward it they remain communicative" (Eco, 2013, pp. 39–40). That is to say, a code is a structure reacting to preconceived premises specifically made for it—a unique code.

The "operational code," a precursor to the geopolitical code, was developed by political scientist Nathan Leites when he undertook his studies on Bolshevism and the decision-making beliefs of the Soviet leaders (Dijkink, 1998). Later George (1969) expanded on the ideas of operational code using Leites's idea to understand value-based decision-making, judgment and space in which individuals—political players—are enclosed in.

The term geopolitical code is only mentioned academically for the very first time by John Lewis Gaddis in Strategies for Containment: A Critical Appraisal of Post-war American National Security Policy during the Cold Warlaunched in 1982. In this book on Gaddis' study of US foreign policy up until the period in which the work was launched, the author, inspired by the ideas of George, clarifies the existence of both strategic and geopolitical codes for US presidential administrations (Gaddis, 1982).

More explicitly, Flint and Taylor (2018) say geopolitical practices produce what Gaddis defines as geopolitical code and provide their own definition for what they understand as "operational codes consisting of a set of political geography assumptions that underlie the development of a country's relations with other countries" (Flint & Taylor, 2018, p. 51).

A State does not have only one geopolitical code at work. All countries have and always will have more than one code in use and these are constantly being modified or deleted. The existing codes even exist simultaneously and in interaction so that they can reach the goal by the person who had established such codes.

Also, according to Flint and Taylor (2018, p. 52) the codes have three non-exclusive scales of action. In fact, the codes and their different scales of action are in constant dialogue. This way a state agent may have a local, regional and global code. We shall see in Table 5.1 the purposes of a geopolitical code in line with its operating scale.

This is why we stated earlier that all states have geopolitical actions and interactions among themselves. I understand that the geopolitical code denotes the basic assumptions to outline a country's decision-making to establish relationships with others, i.e., the construction of its foreign policy and has three distinct levels, but

Table 5.1 Geopolitical codes and their objectives

Local	Regional	Global
The local code entails the assessment of neighboring states. Governments in all countries, however small, should have such a code	They are required for power projection beyond immediate neighbors. All regional powers must have this code	A code used by nations with a global geopolitical sphere of influence

Source Flint and Taylor (2018, p. 52)

in communications, all the state actors are inevitably active on the first scale of the geopolitical code—the local one.

By understanding that the geopolitical code represents a country's main objectives as regards its foreign policy and it is through this code that governments' geopolitical strategies will be put in place, we can advance our comprehension of what we perceive as Chinese geopolitical code and then, its way of operating in Central Asia.

As we stated above, since the end of the Cultural Revolution the Chinese have been pursuing internal and external growth. China's first state policies to develop and promote socio-economic growth were primarily through the so-called Great Leap Forward in Mao Tse-Tung's era and then the process of economic reform during the Deng Xiaoping era.

While China was being restructured internally, the Chinese Communist Party (CCP) also sought to create a favorable external environment for its own interests. One way of establishing dialogue with other countries and broadening its diplomatic relations with other parts of the world was the start of "five rules for a peaceful coexistence: mutual respect for sovereignty and territorial integrity, non-aggression, non-interference in the internal affairs of one country by the other, equality and reciprocal benefit and peaceful coexistence" (Pomar, 2003, pp. 154–155). A peaceful rise as we have pointed out in the introduction to this text. An example of this peaceful rise of Chinese diplomacy, for instance, is the border dispute settlement with Tajikistan. Resolution through diplomacy on Hong Kong and the British is another example.

Clearly, these ground rules in China's diplomatic relationship with other countries is a direct influence of the Century of humiliation, where the territory was divided up by the industrial powers of the nineteenth century under the Manchu dynasty.

No wonder the Chinese political philosophical discourse—creating an influence on the geopolitical code—is built on the idea of gradual evolution and mutual respect among the states of the World System. We can say that Chinese foreign policy thinking has great influence from Confucian philosophy seeing China as a great nation that must lead the World toward great development and cooperation, while seeking to be respectful of the internal affairs of partners and pursuing an alternative world order, one that is more compassionate than the present one led by the West (Dreyer, 2015; Pomar, 2003; Woon, 2018).

After China's integration into the main Western economic and political bodies to some extent acknowledging the CCP's victory over the Chinese nationalists, Beijing has been keen to expand its economic and diplomatic ties with other regions of the world, especially the so-called "Global South". This rapprochement and partnership offer (cooperation) in different areas is indeed the main path that the Chinese had been using to increase its geopolitical sphere of influence.

China has expanded its zone of influence around the world, this also includes Central Asia, by means of cooperative agreements providing technical, financial and social assistance, and by governance mechanisms (Bini, 2016; Paulo, 2014). Further amplification of the Chinese geopolitical code goes along the lines of what Mawdsley (2015) believes to have been the breakdown of the normative order from North to Global South in International Development Cooperation in the period between the twentieth and twenty-first centuries, which according to Duchêne et al. (1973) did little to change the economic relations and the scenario between rich—industrialized—and poor countries—providers of natural resources for the wealthy and less industrialized countries.

So-called emerging countries as average incomes rose at the turn of the century in the so-called Global South—Brazil, China, and India—have been able to reach among themselves and also with other less favored countries of the Global South and establish new economic agreements and relations that partially broke with the hegemonic logic of the North. This means they set up a new pole of power in this geopolitical space.

Thus, it seems quite adequate to us to grasp cooperation as one of, if not the foremost, Chinese geopolitical code. The Chinese have been building their foreign policy around the world through these agreements, establishing new zones of geopolitical influence and promoting great economic leverage in other regions. Nevertheless, to make this geopolitical code operative, China has had to develop two major cooperation platforms: the Shanghai Cooperation Organization and the Belt and Road Initiative.

5.3 The Chinese Geopolitical Code in Central Asia and Its Main Platforms: Shanghai Cooperation Organization and the Belt and Road Initiative

The collaborative relationship between China and the Central Asian countries first started soon after the end of the Soviet Union in 1991 but has deepened since the 1994 visit of the then-Chinese Prime Minister Li Ping. This tour would mark the first visit in many years of a Chinese official to the region, as Central Asians were isolated from the rest of the world while under the Soviet Union.

The first great agreement between the Chinese and some Central Asian states came into existence in 1996. We refer to the Shanghai Five, which was the defensive mechanism for resolving border issues between China and Central Asians and is seen as the genesis of China's first major cooperation platform in the region. Besides China, participants in this mechanism included Russia, Kazakhstan, Kyrgyzstan, and

Tajikistan. In addition to the border issues, such military cooperation has also helped the Chinese to prevent the spread of fundamentalist cells into the Xinjiang region, a crucial economic zone for Beijing and one that is complex due to cultural distinctions with the rest of the country.

Initially, this mechanism was fundamental for the Chinese as it secured their borders and maintained the territory of the nation-state, a legacy they have inherited since the days of the Century of humiliation. Until this moment, the Shanghai Five was a tool for regional stability.

The status of this mechanism changed in 2001 when it gained recognition as an inter-state organization and was renamed as Shanghai Cooperation Organization. From then on, the relationship between the Member States of the organization became based on the Shanghai spirit which is: [a] recognition of the sovereignty of each Member State, [b] non-interference in internal affairs, [c] mutual respect for the cultures of the countries, and [d] win–win cooperation (Cabestan, 2013).

The eight-member organization today—China, India, Kazakhstan, Kyrgyzstan, Russia, Pakistan, Tajikistan, and Uzbekistan—has gradually ceased to be merely an instrument for regional stability and has also become more of a mechanism for regional economic cooperation and integration, given that the US military involvement in the region during the war on terror in Afghanistan led to a closer relationship between the US and the Central Asians.

Soon, the Chinese and also the Russians made use of the Shanghai Cooperation Organization as a way to offset the Western influence in the area, deemed by both countries to be fundamental to their geopolitical strategies—the Russians prioritizing areas, borders and former Soviet infrastructure while the Chinese gave importance to investment in infrastructure (Visentini, 2013; Lewis, 2018).

We may consider that the Shanghai Cooperation Organization is the first platform of the operationalization of the Chinese geopolitical code—cooperation—in Central Asia and its scale of action is regional since it is associated with the most immediate countries bordering China. The Shanghai Cooperation Organization is central to the second platform of operationalization in the Chinese geopolitical code, something we will talk about in the next paragraph.

The second way China puts its geopolitical code into practice in Central Asiatic space is by means of the Belt and Road Initiative. This Chinese cooperation platform is set up on a global scale and has Central Asia as its fundamental place for achieving its objectives.

The Belt and Road Initiative was announced in 2013 by the General Secretary of the Chinese Communist Party Xi Jinping in Kazakhstan. The purpose of the Belt and Road Initiative, according to the Chinese government's plan of action is:

> The Belt and Road Initiative is a systematic project, which should be jointly built through consultation to meet the interests of all, and efforts should be made to integrate the development strategies of the countries along the Belt and Road. The Chinese government has drafted and published the Vision and Actions on Jointly Building Silk Road Economic Belt and twenty-first century Maritime Silk Road to promote the implementation of the Initiative, instill vigor and vitality into the ancient Silk Road, connect Asian, European and African countries more closely and promote mutually beneficial cooperation to a new high and in

new forms. (Action plan on the Belt and Road Initiative, The State Council the People's Republic of China, 2015).

The Belt and Road Initiative is a way of acting out the Chinese geopolitical code and it does so via numerous agreements in infrastructure works created around the World, not just in Central Asia. These works range from a simple irrigation project to building roads, railways, power transmission lines and so forth.

Between 2013 and 2021, the Chinese claim to have signed a memorandum for collaborative agreements with approximately 143 Countries (Belt & Road Portal, 2022). The scope of agreements reached through this platform has given them access to different regions of the world through railways, ports, road connections and logistics efficiency enabling a rapid movement of both the exports exiting China as well as the imports reaching the country.

Most of the infrastructures agreed upon within the context of the Belt and Road Initiative are supported by the Chinese banking system by providing Official Development Finance to fund these works. The Official Development Finance providers are the Chinese policy banks, such as the China Development Bank and the Export and Import Bank of China, which grant credits to the Chinese partners to enable their infrastructure development projects whether it is a port, a factory, a road or an airport. Many of these Chinese-funded works in foreign territories are essential part of the Belt and Road Initiative's economic corridors.

Central Asia matters, in this context, as it is a key region in many ways for the Chinese. China's energy security, for example, depends on the countries of Central Asia. One of the main natural gas pipelines supplying the Chinese starts in the Central Asia—China—Central Asia Pipeline. The pipeline connects extracted natural gas from countries such as Turkmenistan, Uzbekistan, and Kazakhstan to supply the Chinese market. The Central Asian territories host two of the main corridors of the Belt and Road Initiative; New Eurasian Land Bridge and China-Central Asia-West Asia Economic Corridor.

These corridors are important for the Belt and Road Initiative as it connect the Chinese geopolitical code to two important regions for the Chinese economy. The New Eurasian Land Bridge links the Chinese with their main economic trading partners, the Europeans. This corridor, materialized as a railway, begins on the Chinese coast and ends in the Netherlands and Belgium. In Central Asia, this corridor is found on the territory of Kazakhstan through the dry port of Khorgos. The China-Central Asia-West Asia Economic Corridor, which within Central Asia is linked to Kyrgyzstan, Tajikistan, and Uzbekistan, enables the Chinese to connect to ports in the Persian Gulf, a workaround should they have problems crossing the South China Sea.

Finally, we should highlight here the interaction that these platforms for geopolitical code-acting possess. The Shanghai Cooperation Organization, which allows the geopolitical code to act regionally, makes certain stability and integration of China's neighboring countries available, which is needed both for Chinese internal security and for those new projects coming up in the global performance of the Chinese geopolitical code by Belt and Road Initiative.

While one platform is concerned with law and order, regional integration—economic and political—and security issues, the other one is promoting territorial and market integration through countless infrastructure works. We shall see in the following section how these platforms are territorially implemented in the Central Asian region to give a boost to the Chinese geopolitical code.

5.4 The Materialization of the Chinese Geopolitical Code in Central Asian Territories

In the last section of this article, we describe how the Chinese geopolitical code is materialized in the territories of the Central Asian countries. It is important to highlight that we have done this study considering the fact that Chinese cooperation takes place in the Central Asian region through the area of defense—via the Shanghai Cooperation Organization—and infrastructure works through the Belt and Road Initiative.

In this methodology, we have used state-run news agencies from China and Central Asian countries as a source for the defense data. When collecting information on infrastructure works, we took the following databases as useful sources; China AIDDATA, Global Development Policy Center, and China Global Investment Tracker.

In the area of defense, we have found that this cooperation occurs mainly, but not exclusively, through military drills (Table 5.2). In the last 19 years, we have counted 18 joint military actions between the Chinese and the Central Asians. The Central Asian country that has hosted the most military drills is Tajikistan with six drills, followed by Kyrgyzstan and Kazakhstan with three drills each. The one that hosted the least was Turkmenistan, which did not participate in any kind of joint military activity because of its permanent neutral status.

Other than military drills, we can also point to other ways of cooperation in the military sphere such as: (a) donation and trade facilitation of military equipment; (b) military education for Central Asian countries' officers, especially of those not part of the Collective Security Treaty Organization (Marat, 2021); and (c) construction of military infrastructure in vulnerable Central Asian zones such as the border of Afghanistan, as has been recently done in Tajikistan.

Even if it may be at a disadvantage to the Russians in some aspects in the area of defense, partly because of the Soviet military heritage which still exists in the region, the most important thing for the Chinese is not overcoming Russian military leverage in the region. Of course, the Chinese do cherish some influence over geopolitical views in Central Asian military officers who get military education in China, but what Beijing seeks by creating cooperation in the military sphere is to establish safe zones and neighboring armies that are able to at the very least secure vital infrastructure to the Chinese and those in nearby territories.

Table 5.2 Military drills between Central Asian Countries and China between 2002 and 2021

Name	Data	Participants	Country/Location
Exercise-01	10–11 Oct 2002	China and Kyrgyzstan	Borders of the two countries
Coalition-2003	6–12 Aug 2003	Member countries of the CXO	Ucharal, Kazakhstan and Ili, China
East Anti-terror 2006	Unknown	Kazakhstan, Uzbekistan, Tajikistan and Russia	Uzbekistan
Tian-Shan-1-2006	24–26 Aug 2006	China and Kazakhstan	Eastern Region of Almaty and Yining, China
Coordination-2006	22–23 Sept 2006	China and Tajikistan	Kulyab, Tajikistan
The IssykKul Antiterror 2007	29–31 May 2007	Member countries of the CXO	IssykKul, Kyrgyzstan
Peace Mission-2007	9–17 Aug 2007	Member countries of the CXO	Chelyabinsk, Russia and Urumqi, China
Norak-Anti-Terror 2009	April 2009	China, Kazakhstan, Kyrgyzstan, Uzbekistan and Russia	Dushanbe Tajikistan
Peace Mission 2010	9–25 Sept 2010	Countries of the CXO, except Uzbekistan	Kazakhstan
Peace Mission 2012	8–14 June 2012	Countries of the CXO, except Uzbekistan	Tajikistan
Joint exercise	July 2015	China and Tajikistan	Vahdat, Tajikistan
CentrAsia-Antiterror-2015	15–17 Sept 2015	China, Kazakhstan, Russia, Uzbekistan, Tajikistan	Bishkek, Kyrgyzstan
Anti-terrorism exercises	Oct 2016	China and Tajikistan	Gorno-Badakhshan, Tajikistan
Cooperation-2019	May 2019	China and Uzbekistan	Jizzakh, Uzbekistan
Cooperation-2019	Aug 2019	China and Kyrgyzstan	Xinjiang, China

(continued)

Table 5.2 (continued)

Name	Data	Participants	Country/Location
Cooperation-2019	Aug 2019	China and Tajikistan	Gorno-Badakhshan, Tajikistan
TSENTR 2019	19–23 Sept 2019	Russia, China, Kazakhstan, Kyrgyzstan, Tajikistan and Uzbekistan	Orenburg, Russia
Peace Mission 2021	11–25 Sept 2021	Member countries of the CXO	Orenburg, Russia
			Total: 18 joint actions

Source Prepared by the author with information from People's Daily Online[1]; Xinhua News Agency,[2,3]; The SCO Regional Anti-Terrorist Structure[4]; U.S.-China Economic and Security Review Commission[5]; Asia-Plus,[6,7]; 24.kg[8]; ChinaMil[9]; Infoshos.ru[10]; UzDaily[11]

These infrastructures are the projects funded by the Chinese policy banks in the context of the Belt and Road Initiative. Drawing on the consultations we have conducted in databases and the methodology we have applied in the study (i.e., to only count infrastructure works developed with support from Official Development

[1] Backgrounder: Major PLA-related joint anti-terror military exercises and trainings. People's Daily Online. Available in: http://en.people.cn/90001/90776/6325264.html.

[2] China, Kyrgyzstan hold joint counter-terrorism exercise. Xinhua. Available in: http://www.xinhuanet.com/english/2019-08/06/c_138288443.htm.

[3] Uzbekistan, China wrap up joint anti-terror drills. Xinhua. Available in: http://www.xinhuanet.com/english/2019-05/16/c_138064037.htm.

[4] "Peace Mission-2021" exercises among SCO Countries will be held in summer at one of the training grounds of the central military district. The SCO Regional Anti-Terrorist Structure. Available in: https://ecrats.org/en/situation/status/7698.

[5] The Shanghai Cooperation Organization: A Testbed for Chinese Power Projection. U.S.-China Economic and Security Review Commission. Available in: https://www.uscc.gov/sites/default/files/2020-11/Shanghai_Cooperation_Organization-Testbed_for_Chinese_Power_Projection.pdf.

[6] A joint exercise for Tajik and Chinese special operations forces to be held in Vahdat on June 6. Asia Plus. Available in: https://asiaplustj.info/en/news/tajikistan/security/20150604/joint-exercise-tajik-and-chinese-special-operations-forces-be-held-vahdat-june-6.

[7] A joint anti-terror drill for Tajik and Chinese servicemen starts in Gorno Badakhshan tomorrow. Asia PlusAvailable in: https://www.asiaplustj.info/en/news/tajikistan/security/20161019/232235.

[8] SCO security forces exercise cooperation to combat terrorism. 24.kg. Available in: https://24.kg/archive/en/community/177086-news24.html/

[9] China and Tajikistan kicks off joint counter-terrorism exercise. ChinaMil. Available in: http://eng.chinamil.com.cn/view/2019-08/12/content_9587478.htm.

[10] "Norak-Antiterror" is a Military Training for SCO Anti-Terrorist Forces. Infoshos.ru. Available in: http://infoshos.ru/en/?idn=4086.

[11] SCO conducts counter-terrorism exercises in Kyrgyzstan. UzDaily.Available in: https://www.uzdaily.uz/en/post/432.

Finance provided by the policy bank Export–Import Bank of China), we were able to account for at least 27 infrastructure projects throughout Central Asia.

To demonstrate on how Chinese investments are diversified but focused on infrastructure development, we will now list (Table 5.3) the range of projects we could identify within the methodological process proposed in this research:

As we can observe from the table above, Chinese infrastructure projects for Central Asia are manifold. Nevertheless, they are within the same scope, i.e., the creation, and improvement of roads or railways in Tajikistan and Kyrgyzstan, the development or more efficient distribution of resources such as the SCADA system in Uzbekistan. Furthermore, it is important to emphasize that these projects are not randomly funded. They meet demands from China and from the countries where they are located. Energy production in Tajikistan, for example, is not only crucial for the whole country but also for the Chinese who can buy from the Tajiks if needed. The purchase of locomotives for Kazakhstan also meets a similar demand, as well as upgrading road

Table 5.3 Infrastructure projects funded by the Export–Import Bank of China

Country	Projects
Tajikistan	Vahdat-Javan railway section upgrade Dushanbe-Kulma road reconstruction (Kulma Pass) Dushanbe-Kulyab-Khorog-Kulma road Tajikistan Regar—Dushanbe—Khujand 500 kV North–South power transmission, Construction of 220 kV transmission line Lolazor–Khatlon Construction of 500 kV transmission line South–North Construction of 220 kV transmission line Khujand–Ayni Dushanbe No.2 Thermal Power Plant Project Kulma-Dushanbe-Chanak road construction Sughd-500 South-North power transmission line Khujand-Aynipower grid
Kyrgyzstan	Rehabilitation of Osh–Irkeshtam Road Project Osh-Batken-Isfana Road Rehabilitation Rehabilitation of Bishkek–Naryn–Torugart Road Project KyrgyzstanDatka-Keming 500 kV
Uzbekistan	Improvement in the water supply system Purchase of freight locomotives SCADA system Improvement of the coal production and disposal system Improvement in hydroelectricproduction
Kazakhstan	Purchase of locomotives Electrolysis plant Pure oil refinery in Atyrau Modernization of the rail system and border posts on the border with China and Uzbekistan
Turkmenistan	Equipment for drilling and extracting natural gas (Dowletabat-3 gas field)

Source Prepared by the author with data from China AIDDATA (2021), Global Development Policy Center and China Global Investment Tracker (2021)

stretches in Kyrgyzstan, essential for the China-Kyrgyzstan-Uzbekistan International Highway, which connects the Central Asian region to China and Pakistan.

It is worth pointing out in this last paragraph that we have only dealt with one part of the cooperation projects present in Central Asia—the Chinese geopolitical code. This text, in fact, only represents a portion of a long Master's research project carried out by this author over the last two years in the post-graduate program in Geography at the Federal Rural University of Rio de Janeiro in Brazil.

Further institutions, companies, and state or non-state actors are also involved in the functioning of this Chinese geopolitical code in the Central Asian space.

5.5 Conclusions

Despite the Chinese geopolitical code being founded on the discourse of goodwill and peaceful ascension as we mentioned at the beginning of this text, we should not believe that these relations are not exempt of contradictions. As Mawdsley (2015) and Bini (2016) forewarn us, we are not witnessing a redistribution of power from the North to the Global South, since there is a strong asymmetrical power relationship between the actors involved.

While there may be a notion of a future-sharing community in the Chinese conception of the World System, the relations remain asymmetrical and not unencumbered by either side's interests in cooperation agreements. China as the world's second largest economy, manages to have superior economic power to any Central Asian country and this unavoidably creates an imbalance over their relations.

We are not claiming here that China has been promoting 'neo-colonialism'. Very far from it. The Chinese abide by internal affairs and do not promote any kind of intervention or compel the adoption of any economic primer to their partners. The Chinese do not promote retaliation against partners who seek to engage with other axes of power. Uzbekistan, for instance, has just recently concluded a trade agreement with the European Union. Turkmenistan is about to become a full member of the World Trade Organization in which the Chinese are also a member. Kazakhstan until the total withdrawal of US troops from Afghanistan, had allowed the traffic of material and personnel across its territory.

But the contradiction, and arguably the economic relationship between the Chinese and Central Asians, weighs heavily, for example, when Kazakhs, whose population has strong links with the Uighurs, remain silent in the face of Beijing's crackdown in Xinjiang. The same can also be seen in other themes such as the issue of the environment and even during the COVID-19 crisis, when the Kyrgyz wanted to renegotiate their debt with Beijing and failed to obtain a positive return soon after canceling a million-dollar contract to build a logistics hub in the country after popular pressure.

Hence, we do not rule out the possibility that the Chinese have some influence in the Central Asian space through their geopolitical codes. They have been developed to do so and enable the Chinese to attain even more distant geopolitical spaces of influence. In Chinese political philosophy, China must take its place among the world's greatest nations precisely because of its culture, history, economy and ancient society.

Whereas China has a clear geopolitical code for establishing global influence, this does not make it a "neo-imperialist" or "neo-colonizing" country. Chinese are pragmatic and expect their partners to be as such in their agreements and relations. Inter-state geopolitics deals with pragmatic relations in which emotions and personal opinions must be set aside. China has its own interests and therefore shapes its own strategies.

Central Asia, within Beijing's geopolitical chess, is a core region that can never be forfeited to another axis to the zone of influence. Along with the Russians, the Chinese secure Central Asia as a geopolitical ensemble which serves as a kind of belt to prevent the interior of the continent from ever being conquered by other state actors or hegemonic power blocs, providing strategic and fundamental logistical nodes as alternatives to seafaring.

This work is only a snippet of the contemporary geopolitical history of China and Central Asia. Their unfolding is still in full swing. We must remain attentive to the next geopolitical steps taking place within the Asian continent a very important area and one that honors a rich and living culture with a lot to offer but which we know little about.

References

AidData. (2021). https://www.aiddata.org/china. Accessed December 3, 2021.

Belt and Road Portal. (2022). International cooperation. https://eng.yidaiyilu.gov.cn/info/iList.jsp?cat_id=10076&cur_page=2. Accessed January 5, 2022.

Bini, V. (2016). Coperazioneallosviloppo in Africa. Teorie, politiche, pratiche. Mlmezisedizione, Milan.

Cabestan, J. (2013). The Shangai cooperation, Central Asia and the great powers, an introduction one bed, a different dreams? *Asian Survey, 53*(3), 423–435. https://doi.org/10.1525/as.2013.53.3.423

China Global Investment Tracker. (2021). https://www.aei.org/china-global-investment-tracker/. Accessed December 8, 2021.

Dijkink, G. (1998). Geopolitical codes and popular representations. *GeoJournal, 46*, 293–299. https://doi.org/10.1023/A:1006999221884

Dreyer, J. T. (2015). The 'Tianxia Trope': Will China change the international system? *Journal of Contemporary China, 24*, 1015–1031. https://doi.org/10.1080/10670564.2015.1030951

Duchêne, F., Mushakoji, K., & Owen, H. D. (1973). The crisis of international cooperation. [S. l.]: The Trilateral Commission.

Eco, U. (2013). *A estrutura ausente*. EditoraPerspectiva.

Flint, C., & Taylor, P.J. (2018). *Political geography: World-economy, nation-state and locality*. Routledge.

Gaddis, J. L. (1982). *Strategies of containment: A critical appraisal of Postwar American National security policy during the cold war*. Oxford University.

George, A. L. (1969). The "operational code": A neglected approach to the study of political leaders and decision making. *International Studies Quarterly*, *13*(2), 190–222.

Lewis, D. G. (2018). Central Asia: Fractured region, illiberal regionalism. In A. Ohanyan (Ed.), *Russia abroad: Driving regional fracture in post-communist Eurasia and beyond* (pp. 119–134). Georgetown University Press.
Li, X., & Worm, V. (2011). Building China's soft power for a peaceful rise. *Journal of Chinese Political Science, 16*(1), 69–89. https://doi.org/10.1007/s11366-010-9130-2
Marat, E. (2021). China's expanding military education diplomacy in Central Asia. PonarsEurasia. https://www.ponarseurasia.org/chinas-expanding-military-education-diplomacy-central-asia/. Accessed November 14, 2021.
Mawdsley, E. (2015). Development geography: Cooperation, competition and convergence between 'North' and 'South.' *Progress in Human Geography, 41*(1), 108–117. https://doi.org/10.1177/0309132515601776
Medvedev, D. (2008). Interview given to television channels one, Rossia, NTV, Presidential Website, 2008. http://en.kremlin.ru/events/president/transcripts/48301. Accessed April 6, 2021.
Ming, L. (2020). Xi Jinping's vision of a community with a shared future for humankind a revised international order? https://www.nbr.org/publication/xi-jinpings-vision-of-a-community-with-a-shared-future-for-humankind-a-revised-international-order/. Accessed October 16, 2021.
Nye, J. (2012). *O future do poder*. Benvirá.
Ó Tuathail, G. (1999). Understanding critical geopolitics: Geopolitics and risk society. *Journal of Strategic Studies, 22*(2–3), 107–124. https://doi.org/10.1080/01402399908437756
Paulo, S. (2014). International cooperation and development: A conceptual overview. https://papers.ssrn.com/sol3/papers.cfm?abstract_id=2430206#paper-citations-widget. Accessed April 12, 2021.
Pomar, W. (2003). *A Revolução Chinesa*. Editora UNESP.
Raffestin, C. (1993). *Por uma Geografia do Poder*.
The People's Republic of China. (2015). Full text: Action plan on the Belt and Road Initiative, Available via The State Council The People's Republic of China. http://english.www.gov.cn/archive/publications/2015/03/30/content_281475080249035.htm#:~:text=The%20Belt%20and%20Road%20Initiative%20aims%20to%20promote%20the%20connectivity,connectivity%20networks%2C%20and%20realize%20diversified%2C. Accessed January 26, 2021.
Visentini, P. F. (2013). Brics, OCX e Ibas: as coalizões dos emergentes. In P. Visentini et al (Ed.), *BRICS: as potências emergentes: China, Russia, Índia, Brasil e África do Sul.Vozes, Petrópolis: Vozes* (pp. 199–224).
Woon, C. Y. (2018). China's contingencies: Critical geopolitics, Chinese exceptionalism and the uses of history. *Geopolitics, 23*(1), 67–95. https://doi.org/10.1080/14650045.2017.1302429

Jonathan Christian Dias dos Santos is Ph.D. student in Geography at the Federal Rural University of Rio de Janeiro (UFRRJ) and researcher linked to the Laboratory of Economic Geography, Politics and Planning (LAGEP) in the Federal Rural University of Rio de Janeiro (UFRRJ). At present, he has devoted himself to studying China, Russia, Türkiyeand Central Asia in geopolitical and geoeconomic approaches.

André Santos da Rocha holds has been awarded a Ph.D. in Geography from Federal University of Rio de Janeiro (UFRJ) and a Postdoctoral Internship in Public Policies and Human Training from Rio de Janeiro State University (UERJ). He is presently an Associate Professor of the Department of Geography at Federal Rural University of Rio de Janeiro (UFRRJ) and a Permanent Professor of the Master's Program in Geography. He is the coordinator of Laboratory of Economic Geography, Politics and Planning (LAGEP) and works in the areas of Political Geography, Economics and Urban-Regional approaches. He is the editor of Continentes—a journal linked to the post-graduation program in Geography at Federal Rural University of Rio de Janeiro (UFRRJ).

Pablo Ibañez has been awarded with a Ph.D. in Geography from University of São Paulo (USP). He was a visiting researcher at Fudan University, Shanghai (2018–2019). He is an Associate Professor of Geography at the Federal Rural University of Rio de Janeiro (UFRRJ). Currently, he acts with research in the field of Geopolitics dedicating himself to the debate of the Belt and Road Initiative.

Chapter 6
Bridging the Himalayas: Nepal's Evolving Geopolitical Strategy Between India and China

Asis Mistry

Abstract This paper delves into Nepal's intricate diplomatic landscape, focusing on its evolving strategy for managing relations with two regional giants, India and China, and the consequential implications for regional stability. The Himalayan nation has long been at the crossroads of Indo–Chinese strategic interests, prompting a careful balancing act. By dissecting the historical context, this study assesses Nepal's foreign policy trajectory, highlighting pivotal milestones and shifting dynamics. It also examines the drivers behind Nepal's multi-alignment approach, examining the economic, political and security dimensions. Furthermore, the paper digs into Nepal's role in China's Belt and Road Initiative (BRI) and its implications for India-Nepal diplomatic relations. Through a comprehensive analysis of Nepal's geopolitical and diplomatic strategy, this research contributes to a deeper understanding of the complex web of alliances and rivalries in South Asia and offers valuable insights into the delicate task of bridging the Himalayas diplomatically.

Keywords Diplomacy · Nepal · India · China · Himalayas · Multi-alignment

6.1 Introduction

Over the past two and a half decades, Nepal has undergone significant political transitions. It was not until 2015 that Nepal truly embarked on a path of peace and development. After more than 60 years of domestic turmoil and political struggle since the 1960s, the promulgation of a new Constitution in 2015 was seen as a beacon of hope. Today, Nepal ranks 108th among 194 countries in the Global Stability Index. With the increase in stability, Nepal's foreign policy engagement has also intensified, drawing greater international attention. Since Nepal implemented its new Constitution in 2015, the country has seen several high-profile visits from major regional powers. In 2014, Narendra Modi became the first Indian prime minister to

A. Mistry (✉)
Department of Political Science, University of Calcutta, Alipore, Kolkata, West Bengal, India
e-mail: ampolsc@caluniv.ac.in

M. Mishra et al. (eds.), *Contemporary Geopolitical Alignment in the Global South*,
https://doi.org/10.1007/978-3-031-81977-3_6

visit Nepal in 17 years, and in 2019, Xi Jinping became the first Chinese president to visit Nepal in 20 years. As Nepal seeks stability, it is also working to diversify its foreign relations by engaging with nations beyond India and China, including key development partners such as the United States, the United Kingdom and Japan. Now, Nepal has diplomatic relations with 155 countries, 33 diplomatic missions abroad, and hosts 28 resident diplomatic missions, 30 UN offices, and numerous aid missions. Nepal is also the second-highest contributor to UN peacekeeping operations.

However, Nepal's geopolitical significance as a landlocked country nestled amidst the towering peaks of the Himalayas has made it a focal point in the intricate web of diplomatic interactions between India and China. Over the years, Nepal's foreign policy has undergone a substantial transformation as it navigates its relations with these two regional powerhouses. The historical backdrop of Nepal's diplomatic engagement with India and China underscores the complex dynamics that have shaped its strategic choices (Gurung & Joshi, 2020). The strategic sensitivity of Nepal's location has led to an evolving diplomatic strategy aimed at maintaining a subtle equilibrium between these neighbouring giants (Adhikari, 2012). This paper seeks to dissect Nepal's growing diplomatic strategy and multi-alignment approach as it bridges the Himalayas, examining both the motivations driving this strategy and the implications it holds for regional stability.

6.2 Theoretical Framework

The framework for this paper is constructed upon a combination of theoretical perspectives, primarily drawing upon Realism and Constructivism, to comprehensively analyse Nepal's evolving diplomatic strategy between India and China. Realism, as proposed by scholars like Kenneth Waltz, offers a lens through which to examine the power dynamics and security considerations that influence Nepal's diplomatic choices (Waltz, 1979). Realist theory suggests that states primarily seek to maximize their security and survival in an anarchic international system. Nepal's strategic balancing act between the two major powers in its neighbourhood can be understood through this lens. Furthermore, Constructivist theory, as advocated by Alexander Wendt, emphasizes the role of ideas, norms and identities in shaping foreign policy (Wendt, 1992). In the context of Nepal's diplomacy, this framework posits that historical legacies, cultural affinities and shared norms play a significant role in influencing its foreign policy decisions. By amalgamating Realism's focus on power politics with Constructivism's emphasis on ideational factors, there is a possibility to offer a general understanding of Nepal's geopolitical strategy, accounting for both the material and ideational drivers that steer its foreign policy choices. While realism underlines the structural constraints and security dilemmas faced by Nepal in the context of its neighbours, Constructivism sheds light on the historical legacies and cultural affinities that influence its foreign policy. This approach acknowledges that Nepal's foreign policy cannot be solely explained by a single theoretical lens but is a complex interplay of material interests and ideational factors. It recognizes that Nepal's diplomatic strategy is not solely determined by external pressures but is also a product of its unique historical experiences, identity and evolving perceptions of its role in the region.

6.3 Review of Literature

Scholarly discourse on Nepal's diplomatic strategy vis-à-vis India and China has emerged as an area of critical importance, given the region's geopolitical complexity and the country's pivotal position therein. Gurung and Joshi's constructivist analysis underscores the historical underpinnings of Nepal's foreign policy choices, examining the impact of the Sugauli Treaty of 1815 and the subsequent Treaty of Friendship with India in 1950, emphasizing the enduring influence of historical legacies on contemporary diplomatic affairs (Gurung & Joshi, 2020). Building on this historical foundation, Acharya's (Acharya, 2007) work delves into the shifting regional context and underscores the necessity for Nepal to adapt its foreign policy amid the changing dynamics of the Indo-Chinese rivalry (Das, 2022). Furthermore, Das's study provides insights into Nepal's diplomatic balancing act by exploring its economic and political motivations, particularly in light of China's Belt and Road Initiative and India's Neighbourhood First policy (Adhikari, 2012). Pant's analysis extends the discourse by examining the broader geopolitical implications of Nepal's strategic choices, especially in the context of the strategic rivalry between India and China in South Asia (Pant, 2016). These studies collectively highlight the complicated nature of Nepal's diplomatic strategy, situated within a regional landscape characterized by intricate power dynamics, historical legacies and contemporary economic and security interests. However, the existing scholarships also reveal some significant research gaps in the understanding of Nepal's evolving diplomatic strategy between India and China. Firstly, there is a need for a more detailed examination of the specific mechanisms behind Nepal's foreign policy decision-making. Second, the direct impact of historical treaties on Nepal's contemporary diplomatic choices requires deeper investigation. Third, a wide-ranging analysis of the economic dimensions, including the consequences of Nepal's involvement in initiatives like the Belt and Road Initiative, is deficient. Also, the security implications of Nepal's diplomacy have hardly been explored. Besides, a comparative analysis of Nepal's strategies with India and China is absent, and a broader assessment of how Nepal's diplomacy affects regional stability in South Asia remains unaddressed. These underscore the complexity and significance of Nepal's geo-strategic posture in the region, calling for further scholarly inquiry to provide a more holistic understanding of the subject.

6.4 Hypotheses

H1: Nepal's contemporary foreign policy decisions are significantly influenced by historical treaties, such as the Treaty of Friendship with India in 1950, and continue to play a substantial role in shaping its diplomatic strategy.

H2: Economic motivations, particularly Nepal's participation in China's Belt and Road Initiative, have a substantial impact on its geopolitical choices, affecting its alignment and engagement with India and China.

H3: Nepal's geopolitical strategy has notable security implications for the nation, directly impacting its national security and contributing to regional stability or insecurity in South Asia.

H4: Nepal's foreign policy decisions with India and China are not isolated but are interconnected and interdependent, implying that shifts in relations with one nation have consequential effects on relations with the other.

H5: Nepal's diplomatic balancing act between India and China may either enhance regional stability by acting as a bridge or exacerbate regional tensions due to its complex position in the broader geopolitical context of South Asia.

6.5 Methodological Note

The methodological approach adopted for this paper combines qualitative analysis and the case study method. Qualitative analysis is employed to examine and synthesize existing literature, academic research and scholarly perspectives on Nepal's diplomatic strategy between India and China. This involves a comprehensive review of academic articles, policy reports and books to provide a well-rounded understanding of the subject matter. A case study method is adopted to explore and analyse specific instances and events within Nepal's diplomatic history, offering real-world insights into its evolving strategy. This method enables an in-depth examination of key historical treaties, diplomatic engagements and policy shifts, thus enhancing the paper's empirical grounding.

6.6 Discussion

Nepal follows a multi-alignment policy at the core of its foreign policy to achieve autonomy and sovereignty. After the Cold War, China's military and economic capabilities have grown significantly compared to India's. From a realist perspective, maintaining a warm relationship with China is a logical aspect of Nepal's foreign policy. In 2021, China's military budget was USD 293 billion, while India's was only USD 76 billion. Additionally, China's GDP was USD 17.73 trillion compared to India's USD 3.18 trillion (World Bank, 2021a, 2021b). However, as a geographical buffer between two nuclear powers, Nepal is careful to balance its national interests and avoid becoming beholden to either neighbour. Nepal prefers to hedge between its two powerful neighbours (Goh, 2006), China and India, under a multi-alliance policy, which helps avoid exacerbating tensions. Since the Cold War, Nepal has maintained peaceful cooperation with both China and India. The country emphasizes the historical and cultural roots of its non-alignment stance, supporting a multipolar or bipolar approach. Nepal's military is significantly weaker than those of China and India, making military alliances with either country a potential source of conflict. While the concept of non-alignment as an organization has become obsolete since 1991, the policy of multi-alignment remains relevant for Nepal due to the risks associated with having two nuclear-armed neighbours. Since 1991, Nepal has taken a dialectical approach to its foreign policy, acting within its geopolitical constraints to protect its sovereignty and autonomy without entering into "military alliances."

6.6.1 Echoes of History

Nepal's contemporary foreign policy decisions are undeniably influenced by historical treaties, with the Treaty of Friendship with India in 1950 serving as a prominent example. This enduring influence is deeply rooted in the historical context of Nepal-India relations. Since Nepal's unification in 1768, the worldviews of its rulers have shaped its foreign policies through three distinct phases: strict isolation and cautious non-alignment, selective isolation transitioning to active internationalism, and the current pursuit of a new creative balance (Jaiswal & Ranjan, 2022, p. 3).The formation of the Nepali state occurred at a unique juncture in Asian and world history. To the north, the once powerful and advanced Chinese civilization and empire were in decline due to internal issues and external pressures. To the south, the fragmented Mughal rule over traditionally Hindu-Buddhist India was replaced by expansionist Christian Britain, which was consolidating its control over the subcontinent and beyond. Recognizing this evolving regional geostrategic context, the founder of the unified Nepali Hindu monarchy famously described Nepal as "a yam between two boulders" (Jaiswal & Ranjan, 2022, p. 7). He understood the limitations of Nepal's power in its sensitive geostrategic location between two great powers and thus adopted a foreign policy of strict isolation and cautious non-alignment.

The Gorkha-Tibet War of 1788, where Nepal invaded parts of Tibet, marked the beginning of Nepal's evolving worldview, influenced by a weakening China and a growing British presence in the region. This shift led to a more pronounced alignment with Britain during the Rana regime's 104-year rule (1846–1951). The Anglo-Nepali War of 1814–15 and the Treaty of Sugauli in 1816 limited Nepal's diplomacy with Britain, but the 1923 Nepal-Britain Friendship Treaty formalized the return of lost territories and recognized Nepal's sovereignty, paving the way for the appointment of Nepal's envoy to London in 1934 (Jaiswal & Ranjan, 2022, p. 8).

With decolonization sweeping through the Indian subcontinent and the aftershocks of the World Wars, Nepal sought to expand its diplomatic reach. Nepal's assertion of independence was bolstered after India's independence in 1947.The Rana regime was overthrown in 1951, and B. P. Koirala became Nepal's first democratically elected prime minister in 1959. Koirala strengthened ties with China, visited Israel, and established formal diplomatic relations, reflecting his worldview and seeking US support for Nepal's democracy (Jaiswal & Ranjan, 2022, p. 9).King Tribhuvan ended Rana autocracy with India's help and democratic changes led by the Nepali Congress under Koirala. His successor, King Mahendra, opposed Koirala's democratic government but continued his internationalism to protect Nepal's interests amidst changing geopolitics. By the time Britain withdrew, Nepal had established diplomatic relations with major powers, including the US, India, France, China, and the Soviet Union, and joined the UN in 1955.During the Panchayat era (1960–1990), King Birendra proposed Nepal as a Zone of Peace in 1975, maintaining active internationalism. This continued after democracy was restored in 1990 and the transition to a secular federal republic in 2008 (Jaiswal & Ranjan, 2022, p. 10). Today, as Acharya (2007) elucidates, the Treaty of Friendship established a close relationship between the two countries, encompassing economic cooperation, security assurances and diplomatic consultations. While Nepal has sought to assert its sovereignty and independence, this treaty has continued to serve as a reference point in diplomatic discourse, shaping the parameters within which Nepal conducts its foreign affairs. The historical legacy of this treaty has, to a significant extent, defined the contours of Nepal's diplomatic strategy with India, emphasizing the treaty's continued impact on the nation's foreign policy choices.

The Treaty of Friendship's enduring relevance is not confined solely to historical reminiscences but is perceptibly present in contemporary diplomatic interactions. Nepal's continued reliance on this treaty for security cooperation and its role as a foundational document in Nepal–India relations highlight its persistent influence (Gurung & Joshi, 2020). While Nepal has diversified its diplomatic engagements and pursued a multi-alignment approach, this treaty remains an indispensable aspect of its diplomatic toolkit. It underlines the complexities of Nepal's foreign policy as it navigates between historical ties and evolving regional dynamics, substantiating the argument that historical treaties like the Treaty of Friendship with India continue to wield substantial influence over Nepal's contemporary diplomatic choices.

The enduring influence of historical treaties, particularly the Treaty of Friendship with India in 1950, on Nepal's contemporary foreign policy decisions remains a central argument in the analysis of its diplomatic strategy. The Treaty of Friendship, Commerce and Transit, often referred to as the 1950 Treaty, has left a profound impact on the trajectory of Nepal's foreign relations (Das, 2022). This treaty established a close and cooperative relationship between Nepal and India, encompassing economic, political and security dimensions. One of its notable provisions is that it requires Nepal to consult with India on matters of defence and foreign policy. While the treaty has been a subject of debate and discussion over the years, its historical significance is undeniable. It continues to influence Nepal's diplomatic posture with both India and China, as it shapes the parameters within which Nepal conducts its foreign affairs.

The enduring relevance of the 1950 Treaty is evident in the fact that it has not merely remained an artefact of history, but remains an active reference point in contemporary diplomatic discourse. Nepal's foreign policy choices and decisions are often scrutinized and assessed in the context of this treaty (Gurung & Joshi, 2020). For example, whenever Nepal contemplates alterations to its foreign policy or security arrangements, it is met with close attention and reactions from India, reflecting the continued relevance of this historical treaty. Thus, the treaty's lasting legacy underscores the complexities of Nepal's diplomatic strategy and its commitment to maintaining a delicate balance between historical ties and evolving regional dynamics.

6.6.2 Indo–Nepal Open Border Versus Sino–Nepal Natural Border

The most fascinating about the borders and borderlands is that they not only streak the end of a particular state's territory but also allude to the start of a different state. In general, borders are contradictory zones of culture and power, where the twin processes of state centralization and national homogenization are interrupted. But in South Asia, the borders of most of the countries are so penetrable that the ethnic groups residing at border adjacent regions exhibit overlapping cultural ties cutting across the physical borderline of nation-states. Therefore, the study of borders and borderlands is enormously fruitful to acquire knowledge about the interplay between populations dwelling at the margin of nations sharing a common border. The day-to-day practice of trans-nationality, the movement of people and goods across the border helps to transcend the territorial epistemology inherent in state-centrism.

Over the last few decades, Nepal and China have exemplified friendship and mutual understanding. Their relationship deepened after Tibet became part of China, creating a shared boundary of 1439 kms. This was formalized through the Nepal-China Boundary Agreement on March 21, 1960, which replaced the Treaty of Thapathali, recognized China's sovereignty over Tibet, and surrendered all old privileges and rights.

On the other hand, the 1950 Peace and Friendship Treaty established an "open border" between India and Nepal for the free movement of people, enhancing social and trade connectivity. This arrangement has benefited both nations by allowing people-to-people associations for generations which is somehow hard to understand from within the Relist Paradigm. However, their 1880-km border has not been without disputes.

Despite the benefits of the open border, it poses challenges such as terrorism, smuggling and undocumented trade. Issues like the circulation of fake currency and restricted areas requiring special permits further complicate the situation. Problems arose in 1989 when India closed 22 border points, reopening only two later (Mistry,

2020). In 2015, another blockade severely impacted Nepal, highlighting the fragile nature of the border relationship amidst India's "Neighbourhood First" policy.

In November 2019, India's Home Ministry released a new political map showing Jammu and Kashmir and Ladakh as new union territories (ANI, 2020). This followed the abrogation of Articles 370 and 35A in August 2019, which removed the special status of Jammu and Kashmir, prompting the release of a new map. The same map depicted the disputed Kalapani region in the Greater Himalayas as part of India, specifically within the Pithoragarh district of Uttarakhand. Nepal immediately objected, identifying Kalapani as part of its Darchula district in Sudurpashchim province (ANI, 2020). Nepal's Ministry of External Affairs, led by Foreign Minister Pradeep Kumar Gyawali, emphasized protecting Nepal's borders and resolving disputes through diplomatic channels based on historical documents and evidence.

India asserted the map's accuracy. The map and Nepal's objection highlighted unresolved border disputes between the countries, including Kalapani, Lipulekh, Limpiyadhura and Susta. Kalapani holds strategic significance as a tri-junction between India, China and Nepal. The core of the disagreement is the origin of the River Kali. Buddhi Narayan Shrestha, former Director General of Nepal's Land Survey department, identified two main viewpoints. The first, held by Nepali scholars, researchers and the government, claims the river flowing west of Kalapani is the main Kali River, originating at Limpiyadhura or Lipulekh pass within Nepalese territory, thus asserting Kalapani as part of Nepal (Nayak, 2020, p. 356). The second viewpoint, held by the Nepal-India Technical Level Joint Boundary Working Group and Indian embassy officials in Nepal, argues the Kali River originates from a smaller rivulet, Pankhagad, south of Kalapani, making the area part of India.

The River Kali, known by different names at various stretches—Kali in the upper reaches, Mahakali in the middle and Sarjoo or Gogra in the lower—adds to the confusion about its true source and border delineation (Nayek, 2020, p. 357).Nepali maps typically show the Kali River originating about 16 kms northwest of Kalapani at Limpiyadhura in the Zanskar range. An 1827 map, often considered "authentic," notes its publication by James Horst Surgh, Hydrographer to the East India Company. Another map from 1830, updated in 1846, demarcates the river from Limpiyadhura (Nayak 2020, p. 357). Most maps from 1846 to 1880 maintain this geographical positioning, though the river was referred to as "Kuti" and later "Kuti Yangti" during different periods. Most of these maps were created by Nepal, leading India to view them sceptically. After 1880, maps started showing a separate River Kali originating from Lipulekh Pass, leading Nepal to accuse India of "cartographic aggression" as it lost about 310 square kilometres (Nayak, 2020, p. 357).

The debates intensified in the 1990s after Nepal restored democracy. India had stationed troops in Kalapani during the 1962 Indo-China war, viewing it as a strategic defence point. At that time, China also recognized Kalapani as part of India. However, Nepal conducted elections in the area until 1961.India and Nepal have repeatedly discussed Kalapani. In 2000, Prime Minister G. P. Koirala of Nepal and Indian Prime Minister Atal Bihari Vajpayee agreed to conduct a field survey and provide reports with new strip maps. Despite India's refusal to withdraw troops, efforts to resolve the issue continued (Mistry, 2020).

The Sino–Nepal borderland and its surrounding borderland, on the other hand, are characterized by a complex interplay of geographical, historical and geopolitical compulsions. Geographically, the border is situated in the rugged terrain of the Himalayas, presenting unique challenges for demarcation and management. After comprehensive surveys, the Boundary Treaty was finalized on October 5, 1961, based on traditional use, possessions and convenience. Conflicts were resolved using a "give and take" policy: Nepal ceded about 1,836 square kilometres to China, while China gave Nepal 2139 km^2. The watershed principle of the Himalayan range was used for demarcation, and landowners whose pasture lands ended up on the other side of the border could choose their citizenship. Physical demarcation involved resolving disputes in 32 areas, guided by peaceful coexistence principles. Permanent pillars and markers, numbered 1–79 from west to east, were erected on June 21, 1962. These included 48 large, 31 small and 20 offset pillars to protect against natural calamities.

The Nepal-China boundary protocol, signed on January 20, 1963, mandated inspections every five years, and it was renewed three times to repair damaged pillars. Minor conflicts have arisen, such as the disputed placement of pillar 57 in Dolakha district, involving six hectares of land. Another issue concerned the ownership of Mount Everest, which was resolved in 1960 when Chou En-lai confirmed it belonged to Nepal. However, a dispute over Everest's height persists, with China claiming 8844.43 m and Nepal 8848 m. Due to these disputes, particularly over pillar 57, the fourth boundary protocol remains on hold, and boundary talks have stalled since then.

6.6.3 *Nepal's Diplomatic Tightrope*

In 2020, amidst the COVID-19 pandemic, an impending global economic crisis, and destructive floods and landslides, Nepal found itself in political and diplomatic turmoil over its disputed border with India. This dispute exacerbated tensions within the government and reignited debates over Nepal's future relations with India and China.

On May 8, India inaugurated a link road in a contested area near a strategic junction with Nepal and China. Prime Minister KP Sharma Oli, already grappling with numerous domestic challenges, took a strong stance against India, labelling the road as an attack on Nepal's sovereignty. He issued a new map that included the disputed territory within Nepal's borders and quickly passed it through both houses of parliament. Additionally, Oli's ruling Nepal Communist Party (NCP) proposed an amendment to the citizenship law, requiring foreign women marrying Nepali men (mostly Indian) to wait seven years for naturalization. Oli's nationalist approach garnered some support among the Nepalese public but failed to quiet his critics who accused him of poor leadership during a crisis and demanded his resignation (Rafiq, 2020). Some commentators and officials in both India and Nepal accused Oli of using the border dispute to stoke nationalist sentiment and undermine his rivals within the NCP, or of acting under China's influence. Prime Minster Oli responded by claiming that his political rivals within the ruling party were "colluding with India to oust him." His main opponent in the NCP, former Prime Minister Pushpa Kamal Dahal, known alias Prachanda, refuted the accusation as "neither politically correct nor diplomatically appropriate," insisting that he, not India, sought Oli's resignation. While internal rivalries within the NCP have certainly exacerbated Nepal's political crisis, the significant influence of Nepal's two giant neighbours, India and China, in this turmoil is undeniable (Rafiq, 2020).

In May, Indian Army Chief General MM Naravane suggested that Prime Minister Oli's objections to India's road construction were influenced by Beijing. Indian media, particularly outlets aligned with the ruling Bharatiya Janata Party (BJP), criticized Oli for his alleged pro-China and anti-India stance. One channel even warned Oli not to challenge India, on which Nepal "depends heavily." Indian media has intensely covered the activities of the Chinese Ambassador to Nepal, Hou Yanqi, making allegations about her influence over Oli (Rafiq, 2020). While she has been active, meeting numerous senior Nepali political leaders, there is a probability that Beijing is pushing Oli to confront New Delhi. Reports indicate that top Indian intelligence officials have also been engaging with senior Nepali politicians. The Indian media campaign against Oli, however, does not support claims that India is trying to overthrow his government.

On the other hand, Prachanda has had a complex relationship with India. While India facilitated the integration of Nepal's Maoist rebels, led by Prachanda, into the political system in the mid-2000s, it also played a role in his resignation as prime minister in 2009 after he challenged the Nepal Army, a key Indian influence. Interestingly, Prachanda has not used an anti-Beijing stance to gain leverage over Oli. He has recently adopted positions that appear pro-Beijing. For example, while Oli supported the United States Millennium Challenge Corporation's (MCC) Nepal compact, aimed at enhancing electricity availability and reducing transportation costs, Prachanda opposed it. He cited US statements linking the project to the Indo-Pacific Strategy, seen as a China containment policy. In a recent address to the Chinese Communist Party, Prachanda alluded to the MCC project, stating that any developmental assistance inconsistent with Nepal's policy of non-alignment "can't be accepted by any means."

India has long played a paternalistic role in Nepal, providing assistance but also leveraging Nepal's dependence on it for sea access. India, Nepal's largest trade partner, accounted for 65 per cent of Nepal's imports and 57 per cent of its exports in 2017 (Rafiq, 2020). New Delhi has been accused of using connectivity as a tool of influence, having blockaded Nepal three times in the past 30 years, most recently in 2015 after a devastating earthquake. The 1989 blockade was partly due to concerns over Nepal's growing closeness to China. Despite China's increasing military sales to Nepal, India remained the dominant external power. However, the 2015 blockade backfired, stoking anti-India nationalism in Nepal. This coincided with China's Belt and Road Initiative (BRI), through which Beijing pledged significant aid and investment.

China's substantial investments in extending its rail network towards its western and southwestern borders are reshaping the region's economic geography. This expansion provided Nepal with an alternative route to the sea. In 2016, China and Nepal signed a transport agreement granting Nepal access to four Chinese seaports. While India's Calcutta port is geographically closer, Chinese rail service to Tibet offers potentially shorter and more cost-effective maritime access (Rafiq, 2020). In 2017, Nepal officially joined the BRI, signalling to India, a major opponent of the initiative, that it now has alternatives.

Therefore, economic motivations, particularly Nepal's active participation in China's Belt and Road Initiative (BRI), have indeed emerged as influential drivers of its diplomatic choices, significantly impacting its alignment and engagement with both India and China. Nepal's decision to join the BRI reflects its strategic intent to harness economic opportunities and infrastructure development facilitated by Chinese investments (Adhikari, 2012). This economic dimension has opened up new avenues for Nepal's cooperation with China, as the BRI projects aim to enhance connectivity, trade and development in the region. As a result, Nepal's participation in the BRI has led to closer economic ties with China, thus affecting its overall diplomatic alignment.

However, it is indispensable to recognize that Nepal's involvement in the BRI has also introduced complexities into its diplomatic relationships, particularly with India. India has expressed concerns regarding the strategic implications of the BRI, especially when it involves projects in territories disputed between China and India (Pant & Ritika, 2017). Nepal's engagement with the BRI has at times generated apprehensions in India about the potential encroachment of Chinese influence in Nepal. This highlights the nuanced impact of economic motivations on Nepal's diplomatic choices, as it needs to balance its economic interests with India's strategic concerns while pursuing its alignment with China.

India has expressed concerns regarding the strategic implications of the BRI, especially when it involves projects in territories disputed between China and India (Pant & Ritika, 2017). Nepal's engagement with the BRI has, at times, generated apprehensions in India about the potential encroachment of Chinese influence in Nepal. This reveals the nuanced impact of economic motivations on Nepal's diplomatic choices. It underscores the need for Nepal to strike a careful balance between its economic interests, largely associated with the BRI and India's strategic concerns. Thus, while the BRI presents economic opportunities, it also necessitates skilful diplomacy to mitigate potential diplomatic tensions with India.

6.6.4 Nepal's Security Dynamics and Regional Implications

Nepal's diplomatic strategy indeed carries substantial security implications for the nation, exerting direct influence on its national security while also contributing to the broader dynamics of regional stability or insecurity in South Asia. As Nepal engages in its diplomatic balancing act between India and China, the decisions it makes regarding military cooperation, security agreements and alignment with one or the other can have far-reaching consequences (Mistry, 2020). For instance, closer security ties with China raise concerns in India, potentially affecting Nepal's national security by inviting closer scrutiny or even reactions from its southern neighbour. This accentuates the intricate interplay between Nepal's diplomatic choices and its security considerations.

India, the US, and the UK have traditionally been Nepal's main defence partners. Recently, however, China has become one of Nepal's key defence allies. In 1989, Nepal purchased military equipment from China, including anti-aircraft guns, missiles and AK-47 rifles. This ceased with the monarchy's fall in 1990 but resumed briefly during King Gyanendra's reign when Beijing provided around $1 million to combat the Maoist insurgency (Gupta, 2023).Since then, China has consistently provided security assistance to Nepal. This includes $2.6 million in 2008, RMB 20.8 million in 2009, $7.7 million in 2011, and $32.3 million in 2017. In 2017, Nepal and China conducted their first joint military drill, "Sagarmatha Friendship," marking a significant step in their defence cooperation(Gupta, 2023). During Deputy Prime Minister and Défense Minister Ishwor Pokhrel's visit to China in October 2018, an agreement was signed to increase China's military support to Nepal by 50% to enhance disaster management and equip Nepal's UN peacekeeping missions better. Additionally, the People's Liberation Army has expanded war college seats and National Défense Course quotas for Nepal Army officers. This growing collaboration indicates that China is poised to surpass Nepal's traditional defence partners.

Despite previous resolutions of border disputes between Nepal and China, recent reports have indicated Chinese encroachments into Nepalese territory in Humla. On August 21, 2021, the Deuba government established a panel led by Jaya Narayan Acharya, joint secretary at the Ministry of Home Affairs, to investigate boundary issues from LimiLapsa to Hilsa in Humla along the Nepal-China border. The panel identified concerns at Pillars 11 and 12 and recommended resolving these issues through a joint inspection group of experts (Gupta, 2023). Comprising six members from the Nepal Police, Armed Police Force, National Investigation Department and Department of Survey, the panel emphasized the need for bilateral attention to these issues. However, China has denied any such encroachment.

India, on the other hand, India is apprehensive about China's increasing influence in Nepal, historically within India's sphere of influence. With the ratification of the US$500 million Millennium Challenge Corporation (MCC) grant project by the Nepalese parliament in February 2022, the US is expected to take a more proactive role in Nepal (Gupta, 2023). Nepal's perception of the MCC turned sceptical after remarks by the US Acting Deputy Assistant Secretary for South Asia, David J. Ranz, suggesting it as part of the US Indo-Pacific strategy during his 2019 visit to Kathmandu. This perception led Nepal's left parties to view the MCC as potentially exacerbating US-China rivalry in Nepal. Reports and analysts have suggested Chinese involvement behind anti-MCC protests, interpreting the MCC as a strategic move to encircle China under the US Indo-Pacific strategy. This situation sets the stage for increased competition between China and the US in Nepal as well.

So, Nepal's diplomatic posture has some substantial implications for regional stability in South Asia. The delicate nature of Nepal's position in the Indo–Chinese rivalry necessitates careful navigation to prevent exacerbating tensions in the region.

Its role as a bridge between India and China can either facilitate regional stability by promoting dialogue and cooperation or contribute to insecurity if perceived as aligning too closely with one side. This nuanced role makes Nepal's diplomatic strategy pivotal in the broader regional context (Pant, 2016). Thus, the security implications of Nepal's diplomacy extend beyond its borders, influencing the overall geopolitical stability or instability in South Asia. Nepal's choices in terms of security agreements, military cooperation and alignment with either of its powerful neighbours can have profound consequences. For instance, closer security ties with China may raise concerns in India, potentially leading to a revaluation of its relationship with Nepal. This illustrates how Nepal's diplomatic strategy can directly impact its national security by inviting reactions and scrutiny from its immediate neighbours.

Nepal's role as a bridge between India and China, therefore, means that its actions can either facilitate regional stability by promoting dialogue and cooperation or contribute to insecurity if perceived as aligning too closely with one side. Nepal's diplomatic decisions resonate within the larger context of the Indo-Chinese regional rivalry, which has the potential to draw it into geopolitical competition (Pant, 2016). Hence, the interconnectedness of Nepal's diplomacy with India and China is such that shifts in one relationship can trigger a chain reaction, influencing not only its national security but also the overall balance of power and stability in South Asia.

6.6.5 Diplomatic Chessboard

The argument that Nepal's foreign policy decisions with India and China are not isolated but are interconnected and interdependent is substantiated by the intricate web of regional geopolitics in South Asia. Nepal's geographic location and the entwined interests of India and China underscore the interdependence of its diplomatic relations with these two neighbouring giants (Pant, 2016). As Nepal seeks to diversify its diplomatic engagements and navigate its role as a bridge between India and China, any perceptible shift in its relations with one nation can have consequential effects on relations with the other. For instance, Nepal's increased engagement with China under the Belt and Road Initiative (BRI) has at times led to concerns in India about the growing Chinese influence in Nepal, prompting adjustments in India-Nepal relations (Adhikari, 2012). Conversely, when India perceives Nepal's policies as tilting too far towards China, it can lead to recalibrations in its engagement with Nepal. This delicate balancing act underscores the inherent interdependence of Nepal's foreign policy choices with India and China.

China has been urging Nepal to participate in the Global Security Initiative (GSI), which Beijing promotes as a framework to address international conflicts, enhance global security governance, foster collaborative efforts for stability in a volatile world, and promote lasting peace and development globally. However, Nepal's cautious approach to China's proposal to join the GSI reflects concerns about potential security cooperation implications. While the GSI advocates for global peace, it is viewed by some, including officials in Nepal, as a security agenda that may favour China's strategic interests in developing countries. Critics have highlighted risks such as data security, unequal distribution of benefits, lack of transparency, potential economic dependence and debt concerns associated with the initiative. Nepal reportedly indicated its reluctance to join the GSI to China.Moreover, despite statements from the former Chinese ambassador to Nepal suggesting Nepali support for the Global Development and Security Initiative, Nepal has not officially endorsed the initiative. Critics and geostrategic analysts in Nepal have raised concerns over the GSI, particularly questioning China's transparency regarding the specifics of any potential agreement.

Nepal's diplomacy is a microcosm of the broader Indo-Chinese regional rivalry. As Pant (2016) suggests, Nepal finds itself caught in the crossfire of the strategic competition between India and China. Consequently, Nepal's actions and alignments do not occur in isolation; they resonate within the larger context of regional power dynamics. The interconnection between Nepal's diplomacy with India and China is such that shifts in one relationship often trigger a chain reaction, affecting the overall balance of power and stability in South Asia. Therefore, a nuanced understanding of Nepal's foreign policy necessitates acknowledging the intricate interdependence between its relations with these two regional actors.

6.6.6 Diplomatic Nexus

Nepal's diplomatic balancing act between India and China indeed holds the potential to either enhance regional stability as a bridge or exacerbate regional tensions due to its complex position within the broader geopolitical context of South Asia. Nepal's historical role as a buffer state between these two regional giants has often allowed it to serve as a conduit for dialogue and cooperation. By maintaining positive relations with both India and China, Nepal has the opportunity to facilitate regional stability by acting as a diplomatic bridge. As Gurung and Joshi (2020) suggests, Nepal's strategic location has been advantageous in mediating disputes, and its commitment to non-alignment has historically been viewed positively by its neighbours. This bridge-building role can foster cooperation and reduce tensions in the region, thereby enhancing regional stability.

However, Nepal's complex position also presents challenges that could exacerbate regional tensions. The volatile Sino–Indian rivalry has the potential to draw Nepal into geopolitical competition, making its balancing act more precarious (Pant, 2016). Shifts in Nepal's diplomatic strategy, whether perceived as leaning more towards India or China, can evoke reactions from its neighbours and contribute to regional insecurities. Nepal's involvement in initiatives like the Belt and Road Initiative (BRI) and the subsequent concerns raised by India illustrates how Nepal's actions can have repercussions on regional dynamics. Therefore, while Nepal has the potential to enhance regional stability, its complex position also carries the risk of aggravating tensions in the broader South Asian geopolitical context.

The flip side of this diplomatic balancing act is, however, the potential to exacerbate regional tensions. The volatile Sino–Indian rivalry places Nepal in a precarious position, and its diplomatic choices can invite reactions from both sides (Pant, 2016). Shifts in Nepal's alignment, even if unintentional, can be perceived as strategic moves by India or China. This perception can lead to recalibrations in their engagement with Nepal, potentially straining regional dynamics. Thus, while Nepal can enhance regional stability as a diplomatic bridge, its complex role also carries the risk of inadvertently contributing to tensions within the broader South Asian geopolitical context.

6.7 Summary of Arguments

Firstly, the historical treaties, particularly the Treaty of Friendship with India continue to exert a considerable influence on Nepal's diplomatic posture, shaping its relations with both India and China. Secondly, the profound effect of economic motivations, especially Nepal's participation in China's Belt and Road Initiative has introduced new dynamics into Nepal's diplomatic engagements, affecting its alignment with both neighbouring countries. Thirdly, the significant security implications of Nepal's diplomacy, as its choices directly influence its national security while also contributing to regional stability or insecurity in South Asia. Fourthly, the interconnected and interdependent nature of Nepal's diplomatic relations with India and China places it at the centre of the Indo-Chinese rivalry, with shifts in relations with one nation having consequential effects on the other. Finally, Nepal's diplomatic balancing act can either enhance regional stability by acting as a bridge or exacerbate regional tensions due to its complex position within the broader South Asian geopolitical context. The subtle nature of this role underscores the intricacies of Nepal's foreign policy, influencing both regional stability and the larger Indo–Chinese regional rivalry.

Nepal's inclination towards China is a consequence of her intimidating historical experiences vis-à-vis New Delhi and is also an effort to lessen its over-reliance on India. Therefore, Nepal's foreign policy is vigorously straggling nearer strategic integration with Beijing while India is losing its significance in Nepal's foreign and domestic politics (Gurung & Joshi, 2020). Beijing has already toppled India as Nepal's primary economic partner. In 2018, China replaced India's National Hydroelectric Power Corporation over a contract to build Nepal's biggest hydropower plant (Gurung & Joshi, 2020). By contrast, India's FDI assurances haveastounded in the last few years from US$350 million (S$495 million) in 2014–15 to US$18 million (S$25.5 million) in 2015–16. If India lacks resources to suppress China, its over-reliance on soft power diplomacy to maintain ties with Nepal has correspondingly become obsolete. In addition, Chinese cultural diplomacy well sponsored by its economic assistance has led to a cultural transformation in Nepal (Gurung & Joshi, 2020). For example, Mandarin has been made compulsory in many private schools in Nepal. Not only that, events like orchestral concerts and courses on traditional Chinese medicine are organized frequently to promote Chinese culture there.

6.8 Conclusion

The exploration of Nepal's evolving diplomatic strategy between India and China underscores the intricate and dynamic nature of its foreign policy decisions. The analysis of the hypotheses reveals that historical treaties, economic motivations, security considerations and the broader regional context collectively shape Nepal's diplomatic choices. The enduring impact of historical treaties, particularly the Treaty of Friendship with India, continues to resonate in Nepal's relations with both nations. As exemplified by Nepal's participation in China's Belt and Road Initiative, economic motivations introduce new dimensions to its diplomatic engagements. These choices, in turn, have direct implications for Nepal's national security and contribute to regional stability or insecurity. Nepal's role as a bridge between India and China highlights the interconnected and interdependent nature of its relations with these regional powers, resulting in the delicate navigation of a geopolitical chessboard. This nuanced diplomatic balancing act has the potential to either enhance regional stability or exacerbate tensions within the broader South Asian context. It also underscores the intricate tapestry of Nepal's diplomatic landscape and its critical role in shaping regional dynamics in the complex nexus of Indo–Chinese relations.

6.9 Policy Options for Nepal, India and China

Policy options for Nepal, India and China should be driven by a shared commitment to fostering stability, cooperation and prosperity in the region. First and foremost, all three nations should prioritize diplomacy and open channels of communication. Regular high-level dialogues and summit meetings can help address concerns, build trust and promote understanding among the three countries. Additionally, economic cooperation should be encouraged through initiatives like the Belt and Road Initiative, while ensuring transparency, sustainability and fair terms for all parties involved. Simultaneously, diversifying economic partnerships beyond bilateral arrangements can enhance economic stability. Investing in regional integration, such as strengthening the South Asian Association for Regional Cooperation (SAARC), should be explored. Enhanced regional cooperation can lead to shared benefits, greater security and more effective conflict resolution mechanisms. It is also crucial for all three countries to consider the humanitarian and ecological aspects of their policies, ensuring the well-being of the people and the environment. Finally, jointly addressing global challenges, such as climate change and pandemics, can serve as an avenue for cooperation and goodwill.

References

Acharya, A. (2007). The emerging regional architecture of world politics. *World Politics, 59*(4), 629–652. https://doi.org/10.1353/wp.2008.0000

Adhikari, M. (2012). Between the dragon and the elephant: Nepal's neutrality conundrum. *Indian Journal of Asian Affairs, 25*(1/2), 83–97.

ANI. (2020, 19 May). Nepal cabinet approves new mapincorporating Lipulekh, Kalapani, Limpiyadhura, claims itsterritories. Retrieved from https://www.aninews.in/news/world/asia/nepalcabinet-approves-new-map-incorporating-lipulekhkalapani-limpiyadhura-claims-itsterritories20200519011222/. Accessed September 07, 2021.

Das, B. (2022). Relative autonomy of Nepal: Revisiting Nepal's foreign policy toward India and China. *Indian Journal of Asian Affairs, 35*(2), 57–72.

Goh, E. (2006). Understanding "hedging" in Asia-Pacific security. *PacNet, 43*(31).

Gupta, R. (2023). Nepal's geopolitical crossroads: Balancing China, India, and the United States. Asian Society Policy Institute. Retrieved from Nepal's Geopolitical Crossroads: Balancing China, India, and the United States|Asia Society. Accessed January 15, 2024.

Gurung, W. F., & Joshi, Y. (2020). *Indo-Nepal Dynamics vis-à-vis China.* https://www.isas.nus.edu.sg/papers/indo-nepal-dynamics-vis-a-vis-china/

Jaiswal, P., & Ranjan, A. (2022). Nepal's foreign policy in a changing world. Institute of South Asian Studies. Retrieved from https://www.isas.nus.edu.sg/papers/indo-nepaldynamics-vis-a-vis-china/. Accessed August 26, 2021.

Mistry, A. (2020). Indo-Nepal border dispute over Kalapani: A two-level game *39*, 43–55.

Nayak, S. (2020). "India and Nepal's Kalapani border dispute: An explainer", ORF Issue Brief No. 356, April 2020, Observer Research Foundation. Retrieved from https://www.orfonline.org/research/india-and-nepals-kalapani-border-dispute-an-explainer-65354. Accessed September 07, 2023.

Pant, H. V. (2016). Rising China in India's vicinity: A rivalry takes shape in Asia. *Cambridge Review of International Affairs., 29*(2), 364–381.

Pant, H. V., & Ritika, P. (2017). India's response to China's belt and road initiative: A policy in motion. *Asia Policy, 24*(1), 88–95.

Rafiq, A. (2020). Nepal is walking a tightrope between India and China. Opinion. Aljazeera. Retrieved from https://www.aljazeera.com/opinions/2020/7/29/nepal-is-walking-a-tightrope-between-india-and-china. Accessed August 08, 2023.

Waltz, K. N. (1979). *Theory of international politics.* California University Press.

Wendt, A. (1992). Anarchy is what states make of it: The social construction of power politics. *International Organization, 46*(2), 391–425.

World Bank. (2021a). Military expenditure [China and India]. Retrieved from World Bank http://data.worldbank.org/indicator/MS.MIL.XPND.CD?end=2021&locations=CN-IN&name_desc=false&start=2021

World Bank. (2021b). Gross Domestic Product (GDP) [China and India]. Retrieved from World Bank Website: https://data.worldbank.org/indicator/NY.GDP.MKTP.CD

Dr. Asis Mistry is a faculty of the Department of Political Science, University of Calcutta, West Bengal, India. His areas of research interest include Comparative Politics, Ethnic and Nationalist Politics, and Democratic Theory. He has been engaged for over a decade in examining Ethnicity and Nation-building in South Asia. Many of his articles have appeared in several International Journals. The author has also contributed research articles as book sections/ chapters in many edited volumes. Thus far, Dr. Mistry has authored two books, namely, "Ethnic Politics in Nepal: A Theoretical Outlook" and "The Question of Ethnicity: Implication for the Process of Nation-building in Nepal".In addition, he is associated with several reputed International Journals including the Asian Journal of Comparative Politics. Dr. Mistry's two recent publications include

(i) India's Leadership in Shaping South Asia's Political Economy: A Global South Perspective (World Focus), and (ii) Bharat as a Civilisational Nation: Unscrambling the Trends of Tradition and Modernity (MPJSS).

Chapter 7
Current Strategic Moves of India to Accentuate Regional Stability and Global Peace: A Geopolitical Review

Sanjoy Saha

Abstract The chapter systematically examines India's multifaceted pursuit of global leadership through various means, such as diplomatic finesse, economic acumen, military modernization, and technological innovation. India has transformed from a historically passive role to proactive leadership, which is evident in its diversified geopolitical alliances with the United States and European nations and collaborations within the Quadrilateral Security Dialogue, i.e., the Quad framework. The "Act East" and "Neighborhood First" policies emphasize economic diplomacy, focusing on regional connectivity projects such as International North–South Transport Corridor (INSTC) and the Chabahar Port. India is committed to upholding a rule-based order and regional stability, evident in its Indo-Pacific engagement, military modernization, and strategic defense partnerships with the US, Israel, and France. India's investments in cybersecurity, digital infrastructure, and initiatives like "Digital India" position the nation as a key player in shaping global technological norms. The synthesis of these strategic pillars reflects India's vision for a secure, technologically advanced, and influential role on the global stage, emphasizing its determination to actively participate and shape the destiny of an evolving geopolitical landscape throughout the globe.

Keywords Geopolitics · Global ascendancy · Economic diplomacy · Indo–Pacific engagement · Military modernization

7.1 Introduction

Geopolitics is a field of study that explores how geography, political power, and economic resources impact global politics and international relations (Vakulchuk et al., 2020; Yang et al., 2023; Yeşiltaş, 2013). It examines how geographical factors such as location, natural resources, and physical terrain influence a country's political

S. Saha (✉)
Kaliachak College, Sultanganj, Malda, West Bengal, India
e-mail: geosanjoysaha1408@gmail.com

M. Mishra et al. (eds.), *Contemporary Geopolitical Alignment in the Global South*,
https://doi.org/10.1007/978-3-031-81977-3_7

behavior, alliances, and strategic decisions (Noorali & Ahmadi, 2023). Additionally, Geopolitics considers the distribution of power among nations, the impact of historical events, and the dynamics of international competition. The goal is to understand the interplay between geography and politics and how they shape the geopolitical landscape (Cao et al., 2024; Sheng, 2024a, 2024b; Sloan & Gray, 2017).

The chapter "Strategic Horizons: India's Current Geopolitical Moves for Global Ascendancy" examines the influence of geography, political power, and economic resources on India's global positioning. It delves into India's political behavior, alliances, and strategic decisions in international competition by analyzing location, natural resources, and historical events. By understanding how geography and politics interact, the article aims to shed light on the dynamics that shape India's current geopolitical landscape and its aspirations for global dominance. India has emerged as a dynamic force, actively shaping its destiny on the global stage. This chapter will explore the catalysts propelling Indias strategic maneuvers for global ascendancy(Nilsen, 2023). With a rich cultural heritage and a robustly growing economy, India is shedding its historical reticence and assuming an assertive role in international affairs. This shift is driven by strategic alliances, economic prowess, technological innovation, and military modernization (Thakur, 2018).At the heart of Indias geopolitical resurgence lies a deliberate shift from a passive role to one of proactive leadership (Tudoroiu, 2023). India recognizes that its ancient civilization and modern capabilities position it uniquely to play a pivotal role in shaping the course of international events (Cohen, 2004; Wyatt & Papastergiadis, 2023).

.The intricate dance of alliances is a cornerstone of India's strategic playbook. Traditionally, India maintained strong ties with nations like Russia, but the current geopolitical landscape witnesses a diversification of partnerships (Edwards, 2023). The United States, European nations, and collaborations within the Quad framework with Japan and Australia indicate India's intention to broaden its diplomatic footprint (Shaheen & Mu, 2023). This diversification strengthens India's negotiating power and fosters cooperative efforts to address global challenges such as climate change, terrorism, and cyber threats. Economic diplomacy is another pillar supporting India's quest for global ascendancy. The "Act East" and "Neighborhood First" policies underscore India's commitment to fortifying economic ties with nations in the Asia–Pacific region and the Indian subcontinent (Goswami, 2022). Transformative projects like the International North–South Transport Corridor (INSTC) and the development of the Chabahar Port enhance regional connectivity and position India as a regional economic hub, countering the influence of other emerging powers (Qian, 2023).

As part of the Quad alliance, India's active participation in the Indo-Pacific marks a strategic move to safeguard maritime interests and uphold a rules-based order. The Quad's collaborative approach seeks to ensure a free and open Indo-Pacific, countering unilateral attempts to alter the *status quo* (Cannonn & Rossiter, 2022). This engagement aligns with India's broader vision of regional stability and security. Military modernization is a tangible expression of India's commitment to securing its interests and contributing to global peace (Smith et al., 2018). Developing Indigenous defense technologies and strategic partnerships with key players like the United

States, Israel, and France enhances Indias defense capabilities. This bolsters its regional security role and positions India as a reliable ally in addressing emerging security challenges (Charpin et al., 2024).

The digital age has ushered in a new frontier for global influence, and India is keenly aware of the importance of technological prowess. Investments in cybersecurity and digital infrastructure underscore India's commitment to securing critical information and establishing itself as a technology powerhouse on the world stage (Shukla et al., 2023). India's pursuit of global ascendancy is a multifaceted journey encompassing diplomatic finesse, economic acumen, military strength, and technological innovation (Jones & Ball, 2023). India is determined to lead discussions and influence the trajectory of a world in flux. The coming years promise a transformative era for India as it endeavors to participate actively and shape the destiny of an ever-evolving global landscape (Khan, 2023).

This work uniquely explores India's transition from a historically passive role to proactive global leadership. A comprehensive analysis of diplomatic, economic, military, and technological dimensions offers valuable insights into Indias interconnected strategies. By incorporating real-time analysis and expert insights, this work enhances relevance and contributes to a nuanced understanding of India's evolving geopolitical position.

7.2 Study Area

India's quest for global supremacy occurs in the context of its complex geographical landscape. Located at the crossroads of South Asia (Sheng, 2024a, 2024b). Indias strategic positioning holds immense geopolitical significance. The country's diverse geography, which extends from the northern Himalayan Mountain range to the vast Deccan Plateau in the south, is crucial in shaping regional dynamics and resource distribution (Lally, 2023). India's focus on Indo-Pacific engagement underscores its recognition of its maritime geography along the Indian Ocean and the Bay of Bengal, highlighting its role as a naval power. The country's economic diplomacy, as exemplified by the "Act East" and "Neighborhood First" policies, strategically leverages its geographical location to enhance regional connectivity through initiatives such as the International North–South Transport Corridor (INSTC) and the Chabahar Port project (Nisar et al., 2023). India's geopolitical decisions, such as military modernization, are closely linked to its varied topography, encompassing mountainous terrain and vast plains (Dame & Etu, 2024). Furthermore, India's technological advancements in cybersecurity and digital infrastructure consider its population's geographical distribution, demonstrating a comprehensive approach that integrates geography, geopolitics, and geoeconomics in its pursuit of global influence (Popescu, 2021).

7.3 3Methodology

The analysis methodology involved an extensive literature review covering India's geopolitical strategies, economic diplomacy, military modernization, and technological advancements. Scholarly articles, government publications, and reputable media sources were scrutinized to gain insights into historical context and contemporary developments. Information was gathered from various sources, including official government statements, policy documents, and expert analyses. Real-time developments and ongoing initiatives were captured through reputable news sources.

Thematic analysis was used to identify critical themes such as geopolitical maneuvering, economic diplomacy, Indo-Pacific engagement, military modernization, and cybersecurity. The gathered information was organized under each theme to provide a structured and comprehensive overview of India's global ascendancy strategy.

The content was structured into distinct sections to enhance clarity and readability, each focusing on a specific aspect of India's global strategy: geopolitics, economic diplomacy, Indo-Pacific engagement, military modernization, and cybersecurity. The approach developed a cohesive narrative, emphasizing the interplay between India's diplomatic, economic, military, and technological pursuits.

Current events were incorporated into the analysis, integrating information on recent geopolitical developments, economic initiatives, military collaborations, and technological advancements. This approach ensured an up-to-date analysis that balanced historical context with contemporary relevance, offering a nuanced understanding of India's global positioning.

Expert insights and analyses from reputable scholars and geopolitical experts were integrated to add depth and credibility to the discussion. Balancing perspectives allowed for the presentation of a comprehensive view of Indias global aspirations. Cross-referencing from multiple sources was undertaken to ensure accuracy and reliability, and a rigorous fact-checking process was implemented to maintain the integrity of the content.

Critical analysis was a key component, involving an in-depth evaluation of the gathered information and exploring the implications and potential outcomes of India's strategic moves. The methodology aims to provide a comprehensive, well-researched, and nuanced exploration of India's pursuit of global ascendancy across diplomatic, economic, military, and technological dimensions.

7.4 Discussion

7.4.1 Geopolitical Maneuver: Broadening Alliance Horizons

India's current geopolitical strategy focuses on diversifying its alliances, which departs from its historical norms of maintaining steadfast ties with Russia (Hong et al., 2024; Leandro & Oberoi, 2023). Instead, India has been actively broadening its partnerships with major global players such as the United States and various European nations. This strategic shift enhances India's diplomatic leverage and equips it with the collaborative strength to tackle pressing global challenges such as climate change, terrorism, and cybersecurity (Shubayr, 2024; Smith & Kartha, 2018). Russia has traditionally been a crucial partner for India, but recognizing the changing dynamics of global geopolitics, India is recalibrating its approach and extending its diplomatic reach beyond its traditional partnerships (Pant, 2013; Anicetti, 2024). The burgeoning relationship with the United States has become a cornerstone of India's diversification efforts.

The Indo–U.S. partnership is characterized by a shared commitment to democratic values and mutual strategic interests. The two nations collaborate in economic partnerships, defense agreements, and joint initiatives in technological advancements (Jang & Yang, 2024; Samuel, 2007). This alignment strengthens Indias economic resilience and elevates its standing in the global community as a responsible and influential player. India's engagement with European nations has also been revitalized, emphasizing the forging of multifaceted alliances. The European Union and individual countries within the bloc have become critical partners for India in addressing global challenges. This diversification strategy reflects India's adaptability and recognition that a diverse set of alliances enhances its international standing and amplifies its voice on the world stage (Kong, 2024; Singh, 2021).

The strategic pivot toward diversification is not only about expanding diplomatic networks but also a proactive response to the intricate challenges faced by the international community. Climate change, terrorism, and cybersecurity are issues that transcend borders, demanding collaborative solutions. By fostering alliances with a broader array of nations, India proactively addresses these global predicaments. India's alliances with nations at the forefront of environmental stewardship facilitate knowledge exchange, technological collaboration, and concerted efforts to mitigate the impact of climate-related disasters. The fight against terrorism gains added momentum as intelligence-sharing and joint counter-terrorism initiatives become more feasible through diversified partnerships. In cybersecurity, collaborative frameworks with technologically advanced nations enhance India's resilience against cyber threats while contributing to the global discourse on secure digital practices (Dalal et al., 2023; Aarya, 2024).

India's diversification of alliances represents a strategic imperative driven by a nuanced understanding of contemporary global challenges. The nation's adept navigation through a changing geopolitical landscape reflects its diplomatic acumen and commitment to being an active participant in the collective pursuit of global stability and prosperity. As India broadens its alliance horizons, the world witnesses a nation poised to play a pivotal role in shaping the course of international affairs (Thakur & Sharma, 2018; Hassan, 2024).

7.4.2 *Economic Diplomacy and Connectivity: India's Path to Global Influence*

India's pursuit of global ascendancy is rooted in its strategic emphasis on economic diplomacy, which shapes its engagement with nations in the Asia–Pacific region and the Indian subcontinent (Chacko, 2014; Saggu, 2024). India's economic outreach goes beyond mere trade partnerships and aims to solidify relationships, foster development, and position itself strategically on the global economic stage. India's "Act East" policy emphasizes a shift in its economic orientation, extending beyond regional boundaries to forge meaningful ties with countries in the Asia–Pacific region. India is expanding its economic partnerships to leverage the economic potential of dynamic nations in Southeast Asia, enhancing its influence in an area witnessing unprecedented growth and development (Verico, 2024). The "Neighborhood First" policy underscores India's commitment to prioritize economic engagement with neighboring countries in the Indian subcontinent (Aryal, 2022). India aims to create a network of stable and prosperous nations that contribute to regional peace and prosperity by fostering economic cooperation and development within its immediate vicinity. This approach strengthens India's geopolitical standing and promotes economic interdependence, fostering a climate conducive to shared growth (Thommandru et al., 2024).

Strategic infrastructure projects are pivotal in realizing India's economic diplomacy objectives. The International North–South Transport Corridor (INSTC) is a critical trade route that links India, Iran, Russia, and Central Asia, facilitating the movement of goods and bolstering economic ties across the participating nations (Khan et al., 2023). India is positioning itself as a linchpin in regional connectivity by investing in transformative projects like INSTC. Iran's Chabahar port development project enhances India's maritime capabilities and provides a crucial gateway to Central Asia. This initiative counterbalances China's ambitious Belt and Road Initiative (BRI), offering an alternative route that allows India to assert its economic influence in a strategic geopolitical space (Khan et al, 2021). India's economic diplomacy extends beyond regional connectivity projects to actively participating in regional forums and trade agreements that promote a multipolar world order. India aims to create a conducive environment for shared economic growth while mitigating the dominance of any single regional player through collaborations within forums like the South Asian Association for Regional Cooperation (SAARC) and the Bay of Bengal Initiative for Multi-Sectoral Technical and Economic Cooperation (BIMSTEC) (Vivoda, 2024).

India's economic diplomacy, manifested through policies like "Act East" and "Neighborhood First," coupled with strategic connectivity initiatives, exemplifies a multifaceted approach to global ascendancy. By strengthening economic ties, investing in transformative projects, and actively countering regional influences, India positions itself as an economic powerhouse and a key player in shaping the geopolitical landscape. As the nation navigates the complexities of global affairs, its economic diplomacy becomes a linchpin in realizing a vision for a more interconnected, prosperous, and balanced world (Banerjee, 2023).

7.4.3 Indo–Pacific Engagement: Charting India's Course in Maritime Geopolitics

In today's world, the Indo-Pacific region has gained increased significance in global geopolitics. India has placed itself as a critical player in this vast maritime domain, and the Quadrilateral Security Dialogue, commonly known as the Quad, is at the forefront of this engagement. The Quad comprises India, the United States, Japan, and Australia, and it has become a vital platform for addressing shared security concerns and promoting a free and open Indo-Pacific. India's active participation in the Quad is a tangible reflection of its commitment to upholding a rules-based order and fostering regional stability, ensuring that the Indo-Pacific remains free from coercion and unilateral dominance (Dar, 2023; Liu & Jamali, 2021).

India's pursuit of global ascendancy extends beyond diplomatic and economic realms, focusing heavily on military modernization. The augmentation of defense capabilities and advancements in Indigenous defense technologies solidify India's standing as a regional security provider. The nation's collaborations with strategic partners like the United States, Israel, and France have gone beyond traditional alliances, encompassing joint military exercises, technology transfers, and collaborative research and development. These partnerships bolster India's strategic posture and significantly contribute to its ability to address emerging security challenges (Sheng, 2024a, 2024b).

India recognizes the critical link between cybersecurity, technology, and national security in a world increasingly defined by digital interconnectedness. Actively investing in cybersecurity and digital infrastructure, India is positioning itself as a regional technology hub and a significant global player with a stake in shaping technological norms. Safeguarding critical information and infrastructure has become paramount, and India's commitment to developing robust cybersecurity measures reflects its proactive stance in countering cyber threats. By cultivating technological capabilities, India fortifies its national resilience and contributes to the global discourse on secure digital practices. As a participant in shaping global technical norms, India plays a pivotal role in ensuring that the digital landscape remains safe, transparent, and conducive to international cooperation (Shubayr, 2024).

India's multifaceted approach to Indo-Pacific engagement, military modernization, and cybersecurity underscores its strategic pursuit of global ascendancy(Vinod, 2023). Through active participation in forums like the Quad, robust defense collaborations, and investments in technology and cybersecurity, India is carving out a role as a responsible global player capable of addressing the challenges of the contemporary geopolitical landscape. As the nation navigates these complex arenas, its initiatives reflect a commitment to shaping a world order that prioritizes security, stability, and innovation.

7.4.4 *Military Modernization and Defense Partnerships: A Cornerstone of India's Global Ascent*

India recognizes the significant role of military modernization and strategic defense partnerships in achieving global supremacy. Apart from diplomacy and economics, India has strategically invested in bolstering its military capabilities, considering them crucial in cementing its position as a regional security provider. At the heart of this effort lies a commitment to indigenization, a concerted effort to develop advanced defense technologies domestically and reduce dependence on external sources (Ladwig, 2015).

India's emphasis on indigenization is a strategic imperative, fostering the growth of a robust defense industrial base. By developing cutting-edge weapon systems and modern communication infrastructure within the country, India enhances its self-reliance and augments its overall defense capabilities. This indigenization drive positions India as a versatile and agile force, well-equipped to navigate and respond to evolving security challenges with innovation and efficiency (Rossiter & Cannon, 2019).

In addition to fostering Indigenous capabilities, India has formed strategic defense partnerships with key global players, establishing collaborations beyond conventional alliances. Among these partnerships, noteworthy engagements are with the United States, Israel, and France, each contributing distinct strengths to India's military modernization agenda.

The Indo–U.S. defense partnership, exemplified by the Defense Technology and Trade Initiative (DTTI), signifies a deepening collaboration beyond mere bilateral ties. This partnership facilitates the co-development and co-production of defense technologies, allowing for the exchange of technological expertise and innovation. Such collaborations not only strengthen India's defense capabilities but also reinforce its position on the global stage as a technologically advanced security provider (Lee-Brown, 2021).

India's association with Israel, known for its expertise in defense technology, has resulted in collaborative projects covering missile defense systems, unmanned aerial vehicles (UAVs), and cybersecurity. This strategic partnership enhances India's technological prowess and addresses shared security concerns, fostering a cooperative defense environment that bolsters both nations' capabilities.

Similarly, partnerships with France have significantly contributed to India's military modernization and strategic depth. The Rafale fighter aircraft deal and ongoing collaborations in naval technologies demonstrate the depth of the Indo-French defense partnership, positioning India as a formidable force in the Indian Ocean region.

India's commitment to military modernization and strategic defense partnerships is a cornerstone of its global ascent. Through indigenization and collaborations with key international players, India strengthens its regional security posture and contributes to the broader discourse on global security. As the nation continues to evolve as a formidable force on the world stage, these initiatives underscore its vision for a secure and technologically advanced future amidst the complex landscape of global geopolitics.

7.4.5 *Cybersecurity and Technology: Safeguarding India's Digital Frontier*

India has shown astute recognition of the intersection between cybersecurity, technology, and national security in the contemporary landscape, where technology plays a crucial role in global geopolitics. The world's increasing interconnectedness necessitates robust measures to safeguard critical information and infrastructure, and India's proactive investment in cybersecurity and digital infrastructure strengthens its national resilience and positions it as a key player in shaping global technological norms (Lakshminarayanan et al., 2023).

India has adopted a holistic approach to cybersecurity to counter cyber threats' escalating frequency and sophistication. Developing indigenous cybersecurity capabilities is central to this strategy, underscoring India's commitment to securing its digital assets. This approach is further enriched by collaborations with international partners, which facilitate the exchange of expertise and best practices. Establishing dedicated cybersecurity centers and formulating a National Cyber Security Strategy exemplify India's proactive stance in countering cyber threats, ensuring the integrity of critical systems, and safeguarding sensitive data (Sarkar & Shukla, 2023).

Besides cybersecurity, India's technological ambitions aim to become a regional technology hub with a global footprint. The "Digital India" initiative, launched in 2015, is a comprehensive program intended to transform India into a digitally empowered society that has driven technological advancements. India is a leader in the technological landscape, embracing areas such as artificial intelligence, quantum computing, and 5G technology.

The "Digital India" initiative aims to bridge the digital divide and empower citizens by promoting the widespread adoption of digital technologies. The initiative encompasses projects ranging from digital literacy programs to creating a robust digital infrastructure, including high-speed internet connectivity in rural areas. These efforts drive economic growth and empower citizens with access to digital services, education, and employment opportunities (Jain et al., 2023). Advancements in emerging technologies amplify India's global technological standing. India's prowess in artificial intelligence is showcased through initiatives that promote research and development in AI applications across various sectors. India has made significant strides in quantum computing, contributing to the global race for quantum supremacy.

India's active participation in developing and deploying 5G technology positions it at the forefront of the next-generation digital revolution.

India's pursuit of global ascendancy is closely connected to its commitment to military modernization, defense partnerships, cybersecurity, and technological advancement. The convergence of these strategic pillars strengthens India's resilience in an increasingly digital world and underscores its role as a responsible global player capable of navigating the complexities of the contemporary geopolitical landscape. As India continues to advance, these initiatives collectively highlight a vision for a secure, technologically advanced, and strategically influential role on the global stage.

7.5 Conclusion

India's efforts to become a global leader are multifaceted and marked by strategic planning and proactive approaches to international affairs. The country has shifted from its historically passive role to a visionary leadership role, evident in its diplomatic finesse, economic acumen, military modernization, and technological innovation. Geopolitically, India has broadened its alliance horizons, moving away from traditional partnerships and instead focusing on diversifying its alliances to tackle contemporary challenges such as climate change, terrorism, and cybersecurity. The nation's shift in its relationships with the United States and European nations reflects its adaptability and recognition of the importance of diverse alliances in enhancing its international standing. Economically, India's "Act East" and "Neighborhood First" policies, along with transformative projects like INSTC and the development of the Chabahar Port, demonstrate a strategic emphasis on economic diplomacy and regional connectivity. These initiatives position India as a key player in shaping the financial landscape and countering the influence of emerging powers. India's active participation in the Quad and military modernization efforts in the Indo-Pacific highlight its commitment to upholding a rules-based order and ensuring regional stability. The nation's collaborations with the United States, Israel, and France contribute to its role as a regional security provider, addressing emerging security challenges. Technologically, India's investments in cybersecurity, digital infrastructure, and initiatives like the "Digital India" program demonstrate a keen awareness of the importance of technological prowess in the digital age. By actively shaping global technical norms, India positions itself as a significant global specialized player and a key contributor to the discourse on secure digital practices. Military modernization, indigenization, and strategic defense partnerships with key global players further solidify India's position as a regional security provider. These efforts reflect a commitment to addressing evolving security challenges with innovation and efficiency, contributing to the broader discourse on global security.

India's initiatives collectively highlight a vision for a more interconnected, prosperous, and balanced world. As the nation continues its transformative journey, its proactive stance in diplomacy, economics, military affairs, and technology underscores its determination to actively participate and shape the destiny of an ever-evolving global landscape. India's pursuit of global ascendancy is a strategic imperative rooted in its historical legacy, economic potential, and recognition of the interconnected nature of contemporary global challenges.

References

Aarya, A. K. (2024). Global eco-politics: Media discourse and conflicting climate change frameworks. In E. C. Keats & J. C. Patel (Eds.), *Eco-politics and global climate change* (pp. 189–214). Springer.

Anicetti, J. (2024). EU arms collaboration, procurement, and offsets: The impact of the war in Ukraine. *Policy Studies, 45*(3–4), 443–466.

Aryal, S. K. (2022). India's 'neighbourhood first' policy and the Belt & Road Initiative (BRI). *Asian Journal of Comparative Politics, 7*(4), 744–756.

Banerjee, D. (2023). *Competing desires: India's look east and act east policy*. Taylor & Francis.

Cannon, B. J., Rossiter, A. (2022). Locating the quad: Informality, institutional flexibility, and future alignment in the Indo-Pacific. *International Politics*, 1–22. https://doi.org/10.1007/s123400000000

Cao, Q., Feng, Z., Yang, R., & Yang, C. (2024). Conflict and natural resource condition: An examination based on national power heterogeneity. *Resources Policy, 89*, Article 104549.

Chacko, P. (2014). The rise of the Indo-Pacific: Understanding ideational change and continuity in India's foreign policy. *Australian Journal of International Affairs, 68*(4), 433–452.

Charpin, R., London, J., & Vincent, N. (2024). The effect of geopolitical tensions on international research collaborations and its implications for global operations management. *International Journal of Production Economics, 268*, Article 109120.

Cohen, S. P. (2004). *India: Emerging power*. Rowman & Littlefield.

Dalal, D., Kumar, R., & Dass, G. (2023). *Environment and disaster management*. Booksclinic Publishing.

Dame, S., & Etu, N. E. (2024). Assessing the financial impact of climate risk: A Comprehensive review. *International Journal of Multidisciplinary Research and Growth Evaluation, 5*(6), 1–16.

Dar, A. I. (2023). India's Indo-Pacific policy: Unpacking the underpinnings and challenges. *Chinese Political Science Review, 8*(4), 597–617.

Edwards, M. (2023). *Cultivating refuge: A long history of dealing with disaster in North-Central Uganda, 1720–2006 (Doctoral dissertation, Northwestern University).*

Goswami, N. (2022). China and the quest for influence.

Hassan, Q., Algburi, S., Sameen, A. Z., Jaszczur, M., Salman, H. M., Mahmoud, H. A., & Awwad, E. M. (2024). Saudi Arabia energy transition: Assessing the future of green hydrogen in climate change mitigation. *International Journal of Hydrogen Energy, 55*, 124–140.

Hong, P. C., Park, Y. S., Hwang, D. W., & Sepehr, M. J. (2024). A growth theory perspective on the competitive landscape of shipbuilding: a comparative study of Japan, Korea, and China. *Maritime Economics & Logistics*, 1–28.

Iyer, S. (2016). *India-U.S. defense ties: An evolving landscape*. The Indian Express.

Jain, A., Anand, A., & Mahajan, I. (2023). Digital governance. In *Technol policy inclusion intersection ideas public policy, 15.*

Jang, Y., & Yang, J. S. (2024). Environmental policy and the evolution of nuclear trade network: Insights from the European Union. *Structural Change and Economic Dynamics, 68*, 425–432.

Jones, B. M., & Ball, S. J. (Eds.). (2023). *Neoliberalism and education*. Taylor & Francis.

Khan, A., Singh, S., Singh, B., & Kaur, A. (2023). India's quest for energy security in Eurasia: A conceptual assessment of energy diplomacy. In *India's energy diplomacy in Eurasia: Geopolitical and geo-economic perspectives* (pp. 111–140). Springer Nature Singapore.

Khan, M. S. (2023). *The mediating dragon in the graveyard of empires: An analysis of China's conflict mediation approach.*

Khan, Z., Khan, Z., Khan, Z., & Khan, Z. (2021). India's maritime strategic outreach in the Indian Ocean region: Power projection and escalation dominance. In *India's evolving deterrent force posturing South Asia: Temptation pre-emptive strikes, power projection, escalation dominance* (pp. 85–136).

Khanna, R. (2019). *China's expanding influence in the Indian Ocean.* The Hindu.

Kong, B. (2024). *SARS STORIES: Affect and archive of the 2003 pandemic.* Duke University Press.

Ladwig, W. C., III. (2015). Indian military modernization and conventional deterrence in South Asia. *Journal of Strategic Studies, 38*(5), 729–772.

Lakshminarayanan, R., & Yepremyan, T. (2023). Armenia-India partnership: Geopolitical and geo-economic implications in the Eurasian context. *Asia Europe Journal, 21*(1), 81–100.

Lally, J. (2023). *India and the early modern world.* Taylor & Francis.

Leandro, F. J., & Oberoi, R. (2023). Ukraine geopolitical European flashpoints vis-à-vis India and China: From ambivalence to strategic engagement. *Janus. Net, 14*(1), 96–116.

Lee-Brown, T. T. (2021). T*he rise and strategic significance of the Indo-Pacific.*

Liu, H., & Jamali, A. B. (2021). India's Indo-Pacific strategy: A pragmatic balancing between the United States and China. *Pacific Focus, 36*(1), 5–39.

Nilsen, A. G. (2023). *Politics from below: Essays on subalternity and Resistance in India.* Taylor & Francis.

Nisar, R. D., Kaleem, M., Abbas, A., & Haider, W. (2023). Navigating apropos midst Kautilya's Mandala's ploy and contemporary India's international relations plays: An analysis. *Pakistan Journal of Humanities and Social Sciences, 11*(4), 4228–4242.

Noorali, H., & Ahmadi, S. A. (2023). Iran's new geopolitics: Heartland of the world's corridors. *GeoJournal, 88*(2), 1889–1904.

Pant, H. V. (2013). India-Russia ties and India's strategic culture: Dominance of a realist worldview. *India Review, 12*(1), 1–19.

Popescu, A. I. C. (2021). The geopolitical impact of the emerging technologies. *Bulletin of "carol I" National Defence University, 04*, 7–21.

Qian, X. (2023). Structural factors of international relations that affect China-Pakistan economic corridor. In *The political economy of the China-Pakistan economic corridor* (pp 21–46). Springer Nature Singapore.

Rossiter, A., & Cannon, B. J. (2019). Making arms in India? Examining New Delhi's renewed drive for defense-industrial indigenization. *Defense Studies, 19*(4), 353–372.

Saggu, D. M. (Ed.). (2024). India at G20: governance & international organizations. *Interdisciplinary Institute of Human Security & Governance.*

Samuel, C. (2007). Indo-US defense cooperation and the emerging strategic relationship. *Strategic Analysis, 31*(2), 209–236.

Sarkar, G., & Shukla, S. K. (2023). Behavioral analysis of cybercrime: Paving the way for effective policing strategies. *Journal of Economic Criminology*, 100034.

Shaheen, N., & Mu, R. (2023). Elevated strategic partnership between India and Japan in the context of the rising power of China. *India Review, 22*(4), 433–462.

Sheng, E. L. (2024). A review on the geopolitics of china: from perspectives of history and theories. In *From colonial seaports to modern coastal cities: The Bohai economic rim and China's rise as a maritime power* (pp. 1–28). Springer Nature Singapore.

Sheng, E. L. (2024). China at the geopolitical crossroads: The construction of sea power. In *From colonial seaports to modern coastal cities: The Bohai economic rim and China's rise as a maritime power* (pp. 29–58). Springer Nature Singapore.

Shubayr, N. (2024). Nuclear security measures: A review of selected emerging technologies and strategies. *Journal of Radiation Research and Applied Science, 17*(1), Article 100814.

Shukla, S., Bisht, K., Tiwari, K., & Bashir, S. (2023) Comparative study of the global data economy. In *Data economy in the digital age* (pp. 63–86). Springer Nature Singapore.
Singh, M. (2021). Multilateralism in a changing global order: Prospects for India–EU cooperation. In *EU-India relations: The strategic partnership in the light of the European Union global strategy* (pp 275–290).
Sloan, G., & Gray, C. S. (2017). *Geopolitics, geography and strategic history*. Routledge.
Smith, P., & Kartha, T. (2018). Strategic partners or an emerging alliance? India and the United States in an era of global power transition. *Computer Strategy, 37*(5), 442–459.
Thakur, R. (2018). How representative is BRICS? In *Emerging powers and the UN* (pp. 43–60). Routledge.
Thakur, R., & Sharma, A. (2018). India in Australia's strategic framing in the Indo-Pacific. *Strategic Analysis, 42*(2), 69–83.
Thommandru, A., & Maratovich, F. F. (2024). Leveraging new emerging technologies to enhance the anticorruption arsenal: Assimilating India's experiential insights into Uzbekistan's Criminoverse. *Multidisciplinary Review, 7*(3), 2024058–2024058.
Tudoroiu, T. (2023). *The geopolitics of China's belt and road initiative*. Taylor & Francis.
Vakulchuk, R., Overland, I., & Scholten, D. (2020). Renewable energy and geopolitics: A review. *Renewable and Sustainable Energy Reviews, 122*, Article 109547.
Verico, K. (2024). *Indonesia's international economic strategies*. Springer Nature.
Verma, N. (2019). *Strategic realignments in the Indo-Pacific*. The Indian Express.
Vinod Kumar, T. M. (2023). The smart megacity system of Indo-Pacific: Emerging architecture and megacities studies. In *Indo-Pacific smart megacity system: Emerging architecture and megacity studies* (pp. 3–159). Springer Nature Singapore.
Vivoda, V. (2024). The geopolitics of South Korea's critical minerals supply chains. *Resource Policy, 81*, 103427.
Vivoda, V., Matthews, R., & McGregor, N. (2024). A critical minerals perspective on the emergence of geopolitical trade blocs. *Resources Policy, 89*, 104587.
Wyatt, D., & Papastergiadis, N. (2023). 'Meeting differently': Indian Independence Day celebrations in the digital diaspora. *City, Culture and Society, 34*, 100521.
Wong, J. (2024). *The new geopolitics of Southeast Asia*. Taylor & Francis.
Yang, Y., Xia, S., & Qian, X. (2023). Geopolitics of the energy transition. *Journal of Geographical Sciences, 33*(4), 683–704.
Yeşiltaş, M. (2013). The transformation of the geopolitical vision in Turkish foreign policy. *Turkish Studies, 14*(4), 661–687.

Sanjoy Saha is an accomplished academician with an M.A. in Geography, where he was awarded a Gold Medal and a Ph.D. specializing in Fluvial Geomorphology. He serves as an Assistant Professor and the Head of the Department of Geography at Kaliachak College, SultanganjMalda, West Bengal, India. Dr. Saha has made significant contributions to the field of geography, having over 10 research publications and 4 book chapters in prestigious international journals such as Elsevier and Springer. Some of his notable research articles include: "*Investigation of the watershed hydro-morphologic characteristics through the morphometric analysis: A study on Rayeng basin in Darjeeling Himalaya*", published in Environmental Challenges (2022); "*Searching missing links between hereditary vocation, community development, and well-being: A case study of snake charmer community (West Bengal, India)*", published in SN Social Sciences (2022) and "*Prioritizing Sub watersheds in Baitarani River Basin, Eastern India*" published in Organization Development Journal (2024). Currently, Dr. Saha is focusing his research on contemporary geopolitical issues, expanding his expertise and contributions to pressing global challenges.

Chapter 8
New Geopolitical Strategy of Maghreb Africa in the New Global Order

Suresh Kumar

Abstract New Geopolitics refers to the post-1990 period of globalization encompasses Maghreb Africa broadly refers to North Africa having regional arrangements and new alignments probing their interests and suit new world order having a spectrum of actors and entities with a wide range of purposes, structures, and obviously geographical coverage. Arab Spring cocktails democracy, aristocracy, and liberal market mechanisms in the name of general policy reforms and marks their presence in the emerging market economy. The strategy of union merely referred to as regional integration increases the economic scales, diversifying their economic production in the combination of the individual Maghreb state's changed prerequisite vis-a-vis market competitiveness. Promoting Maghreb's good governance by reforming its bureaucratic setup, and decentralizing taxation and revenue sharing at the provincial and local level initiates a step in understanding the new geopolitical functioning. The unwanted and indubitably entry of ICT in Maghreb neither sojourned the youth to understand the new realities happening around the world nor did the European influence want to control it in Maghreb. A new Geopolitical strategy has been witnessed emerging of Civil Society organizations, NGOs, Online youth groups, religious leaders, faith communities, and faith-based organizations, and awareness of labor rights in Maghreb Africa on the one hand, and mass movements against rampant corruption has set the trail to new global order on the other hand. This chapter recommends the result of the peaceful mass movement in Maghreb Africa has successfully established the rule of law in different states and promoted good governance.

Keywords Arab spring · Electoral democracy · Internet · Youth · Market economy

S. Kumar (✉)
Department of African Studies, Faculty of Social Sciences, University of Delhi, Delhi, India
e-mail: skafrica2003@gmail.com

M. Mishra et al. (eds.), *Contemporary Geopolitical Alignment in the Global South*,
https://doi.org/10.1007/978-3-031-81977-3_8

8.1 Introduction

The common Islamic religious identity of Algeria, Burkina Faso, Cameroon, Chad, Egypt, Guinea, Libya, Mali, Mauritania, Morocco, Niger, Nigeria, Senegal, Sudan, and Tunisia refers to Maghreb and Sahil Africa. Maghreb Africa refers to Algeria, Libya, Mauritania, Morocco, and Tunisia and extends to Egypt and Sudan being a part of North Africa. Geopolitically, the physical geography of Maghreb has an abundance of natural resources, different tribal groups living together in this region, and similar demography determines its social and economic environment. It plays a very important role in the geopolitics of world history because Italy, France, and other countries have their colonies in that region and they uphold their supremacy in these areas. Maghreb Africa characterized by a combination of high unemployment, a growing young population, and increasing inequality is more prone to instability. Geopolitical study of world order and Maghreb conflicts from a geographical perspective and is preoccupied with explaining the behavior of states by looking at their physical attributes. The geographical position of a country (its regional context specifically) determines the environment within which it will trade; its demography will have a bearing on economic and social development, and the natural resources it possesses or lacks will have an effect on its pattern of economic development. This context inevitably determines how countries will formulate their foreign policies in order to maximize their security and economic advantages.

The people living in Egypt, Morocco, and Tunisia are largely engaged in agricultural activities and subsided with manufacturing activities, trade, and tourism. Algeria, Libya, and Sudan are rich in oil and gas reserves and Morocco has the largest reserves of Phosphates. The post-Covid pandemic has been the cause of a sharp decline in tourism and foreign investment across the region. The Cold War experience of world order encompassed both the legal domain of equality and inequality Geopolitically. This inequality has been unpaid with the African exploitation of raw materials and resources for their industrial development. Global Geopolitics has emerged the various growth centers in the world in which the developing world had been completely ignored or isolated during the Cold War period but the Eurocentric and American approach has been shifted toward the southern global including Africa continent as it has abundance of natural resources for the present and future requirement of the society. Geopolitics of mixing diplomacy and military power of Cold War days and building strategy cannot work in the contemporary liberal market globalization. New global order analytically refers to the arrangement of power and authority, which provides the framework for the conduct of diplomacy globally and the preferred arrangement of power and authority prescriptively associated with international peace, human rights, economic growth and equity, and a sustainable environment. The new Geopolitical environment in North Africa highlights the result of the Arab Spring emphasizing the importance of the physical nature of land and water (Seapower) as a part of their geostrategic alignments. Specifically, the fringe inland of the Mediterranean Sea in North Africa tied by a land bridge to the Near East, separated from Europe (Fig. 8.1) by only very narrow bands of water, has contributed to the advance of civilization over many centuries. Maghreb Africa stands as a neighborhood with regards to Western Europe on the Mediterranean route.

The diversity of endowments in the Maghreb region represents an important opportunity for further development through integration. Morocco, Tunisia, and Egypt have strong private sector and diversified production sectors. Algeria and Libya have a surplus of capital and large markets for goods and services and potential employment opportunities in this region. The opportunities for mutual benefits through cross-border investment and trade between these two groups of countries are visible. The countries on the southern bank of the Mediterranean basin always feel the pressure of the USA's global power and the European Union (France in particular) in terms of economic and political arrangements as compared to the countries on the northern bank in North Africa. French naval base at Bizerte in Tunisia Mers-el-Kebir in Algeria and American naval and air base in Morocco signifies the debate on the

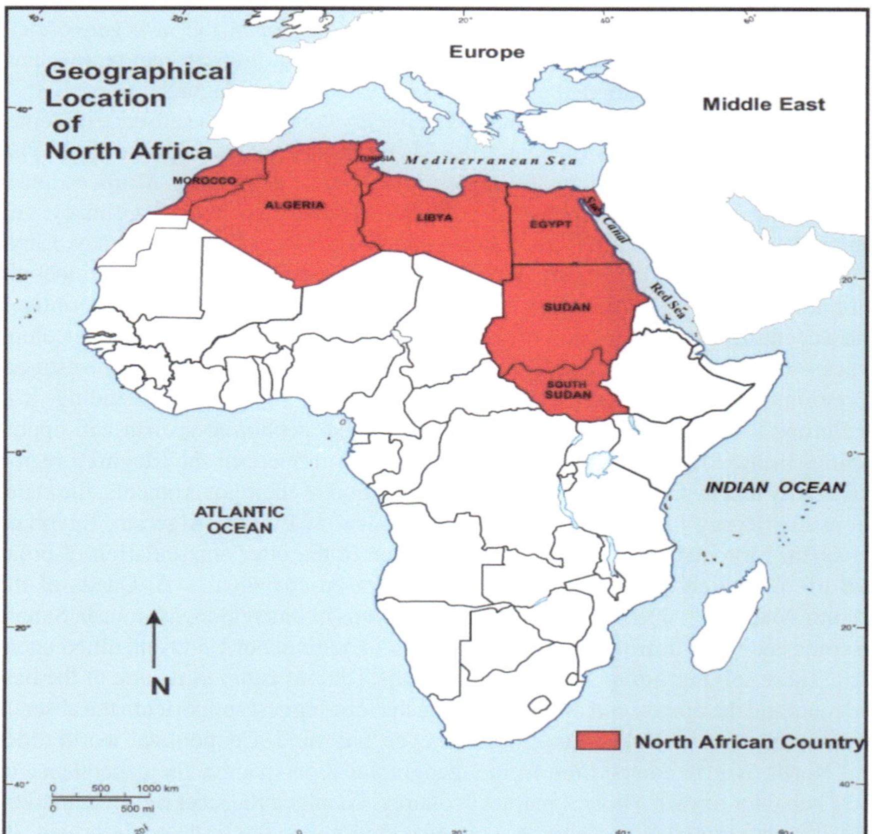

Fig. 8.1 Maghreb and extended North Africa. *Source* http://images.app.goo.gl/CIAGgpwCt5juc983A; accessed on 23.01.2024

Mediterranean basin. Broadly, Geopolitical Maghreb had witnessed constitutional amendments in Algeria, the death of President Muammar Qaddafi and a change of government in Libya, a change of regime in Mauritania, Constitutional amendments and financial powers to villages or decentralization in Morocco, Tunisian peaceful revolution, democratic change in Egypt and division of Sudan into Sudan and South Sudan in the post-1990s period onwards. Arab Spring has raised the voices against the prominent issues of unemployment, rampant corruption, demand for economic infrastructure and development, industrialization, better education, health and sanitation, and social reforms had the common denominators demanded for market-led economic liberalization in the accommodation of new global order.

This accommodation further emphasizes the globally recognized legal norms, liberal principles of governance, and economic liberalization as the foundation of the democratically elected government. The hegemony of the Western World cannot dictate the Maghreb countries as the Geopolitical understanding has widely broadened its sphere in the liberal market economy and particularly in the post-pandemic phase. The national interests always have prime concern in the new geopolitical discourse and neither entertain any West-supported authoritarian regime nor have any place for the West-aligned practice of repression in the Maghreb.

This supportive legitimacy although is a slippery concept in the absence of democratic electoral national politics and the Maghreb has mixed response to it. This mixed response does not guarantee the result of genuine democracy. Morocco under Muhammad VI has institutionalized political pluralism, bolstered legitimacy, and undertaken constitutional reform; Tunisia underpins institutional democracy; Libya is only in the beginning stage; Algeria possesses a certain degree of democratic formation; Egypt institutionalizes the democratically elected form of government; and Sudan still faces the turmoil of democratization. Even the result of the Barcelona process of the mid-1990s, the EU's Neighborhood Policy and colossal investment opportunity have always acted with mutual suspicion and misunderstanding. It is in Europe's own interest to assist in promoting real economic growth and opportunities in the Maghreb region. Political Islam has influenced the Maghreb region differently and in spite of the political legitimization of their governments, the state-sponsored legacy of violence and authoritarianism of Libyan, Algerian, Egyptian, Tunisian, Moroccan, and Sudanese has resulted in the utter emasculation of political life ultimately provide sympathy to jihadist groups such as Al-Qaeda of the Islamic Maghreb (AQIM) and other splinter groups in this region. Moroccan Sahara as contested by Polisario is still hanging in the melodrama of UNO and allied agencies. These splinter groups ultimately made the Tindouf camp as its one of the best hideouts and the open secret of Algeria being the lone biggest supporter in the absence of Erstwhile Soviet Russia has closed its eyes and mind. Geopolitical world order and North African competition from a geographical perspective the dependence of their behavior toward a liberal market economy. Gianluca Salsecci rightly remarked that 'the geographical position of a country determines the trade environment, its social and economic development, and the natural resources it possesses or lacks will have an effect on its pattern of economic development. This context inevitably determines how countries will formulate their foreign policies in order to maximize their security and economic advantages. In the case of the Mediterranean basin, the countries on the southern bank tend to feel the pressure of the international system more than their counterparts on the northern bank, as the latter are the ones determining how the system will operate' (Salsecci, 2010: net). Ultimately, the Maghreb region did not achieve its objective of regional integration and was hindered in making a common symphony either with the African Union (AU) or with the Arab Maghreb Union. Morocco and Algeria are suffering from the Marxist virus of the Cold War and ultimately, the whole region is suffering.

Youth as a human resource dominates the demographic dividend in the Maghreb region and should be seen as an affirmative source of energy for economic development. Geostrategic resources such as hydrocarbon, phosphate, oil, and gas straddling the Mediterranean Sea, Atlantic Ocean, and Straits of Gibraltar's closest proximity to Europe have engrossed the global powers for investment. This dividend comprises more than 50 percent under the age of thirty and demands for an accountable government to talk about minimizing alienation, pessimism, and migration to Europe. "The sheer volume of illegal migrants transiting through North Africa on their way to Europe is not only a humanitarian disaster but it is straining infrastructure, budget, and security in Europe. This challenges the political stability in Europe and NATO's credibility as an effective collective security institution" (Antonio, 2016: net). The post-pandemic environment in the Maghreb region faces the consequences of illegal migration of large-scale youth to European countries, which has become a sort of international media news and no concrete step or action has been taken by the Maghreb governments in this regard. The extraction of natural resources and the industrial development side-by-side will maintain Maghreb's Geopolitical understanding and its significance on the one hand and Maghreb's approach against all sorts of terrorism, no more international funding catering to so-called refugee status and support to Tindouf-based terrorism and resolving the Moroccan Sahara fatality will work for regional integration in the new global order on the other hand.

8.1.1 Geoponic Resources and Global Order

The emerging market leaders such as China, India, Japan, and Brazil in the post-1990 period are looking for alternative energy resources leaving the dependency on the Middle East region. Maghreb North Africa is the substantive alternative for these new economic powers. As a result, Egypt, Sudan, Libya, Algeria, Tunisia, and Morocco have become lucrative sources of oil and gas attracting the global market for their investment in this area mainly from India and China. The global energy demand rises continuously despite its use of renewable resources (such as solar, wind, and hydel energy). 'Global coal emissions grew the most of any fuel over the 2005–2017 period at 2.1 annually. Natura-related emissions grew by 2.0% annually and petroleum-related emissions grew by 1.1 percent annually (EIA Report, 2018) and it was boosted by strong economic growth with oil, gas, and coal meeting most of the demand. With about 70% of global energy demand and unstable conditions in some

Middle Eastern countries being the largest supplier of oil and gas, the market looks toward an alternative supplier and Maghreb North Africa is the most suitable option for them. It is widely known that the Maghreb region oil has low Sulfur content and is widely approachable to Europe and other countries. This quality of oil qualifies the parameter of less emission of Carbon gases comes under sustainable oil consumption and is preferred as compared to high Sulfur content oil.

Figure 8.2 explains the mineral richness of the Maghreb region having petroleum, phosphate, iron ore, silver, zinc, copper, lead, manganese, gypsum, uranium, coal, and cobalt. Petroleum and natural gas exports provide most of Libya's revenues and the rest of the region contains sizeable mineral deposits of these resources. A new global order should focus on peace and security in this region, once it is established, the exclusive development of mineral resources led to industrialization will approach with an affirmative response. Maghreb governments need to pace with this new global order and implement pro-people policies to keep away internal conflicts, and any kind of civil unrest leading to political instability and chaos. 'The experience of the World Policy Conference of 2023 explains the present Sahel region's volatile political environment, the best place for the international terrorist's hideout, is the persistent challenge and a danger to the Maghreb region. The change in the callous attitude and eradicating the hangover of the Cold War days of Maghreb governments is the prerequisite for peace and development. Even the temporary industrial policy arrangement of the Maghreb region cannot resolve the long-lasting problem of unemployment, migration of all sorts, and broadly social security' (Suresh, 2024). The government's channelization of procurement of geoponic mineral resources will remain intact only with a strong voice and policies against international or any supported splinter terrorism and policy implementation irrespective of any democratic elected political party or parties will form the government. Any ill will toward democratization will provide an illicit option of a black-market economy of selling oil and other precious mineral resources (the active terrorist groups such as Boko Haram, Polisario, al Qaeda, etc., in this region widely) in the market liberalization. Different peacekeeping groups of UNO and AU experienced the extremist new hideouts in Libya, Algeria, and Egypt and explain that 'An Islamic states outpost in the southern Mediterranean and about hundred miles from European shore has become a regular threat for the American and European security and economic interests, puts neighboring states at risk and is unlikely that a stable political order and economic development will emerge in North African countries in near future. The Islamic states command about 3000 fighters in Libya and half of those are believed to be Tunisian, Moroccans, Sudanese, and others (Westpoint, 2024:net) is the persistent threat to the Maghreb region and the present Israel and Hamsa war is the result of illegal trading of weapons through the Sinai Peninsula in Egypt. The international terror took advantage of this security vacuum and plentiful arms are in the pools of young idle men in the name of Jihad. Besides, the large ungoverned territories, areas with no or minimal population, lack of adequate infrastructure, or no infrastructure at all still using the traditional mode of transportation and other means of communication is the best fertile land for terror breeding.

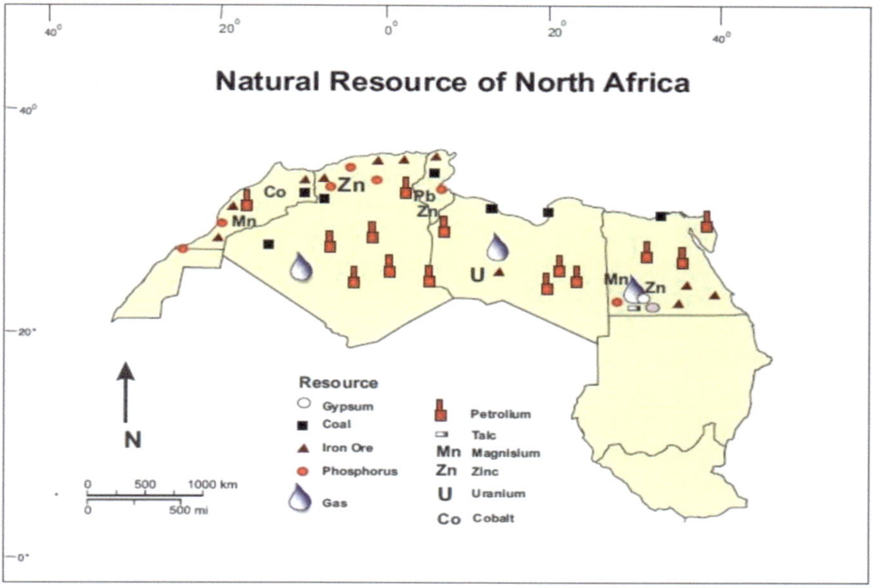

Fig. 8.2 Maghreb region geoponic resources. *Source* http://images.app.goo.gl/wgQKsZgwiiepFAhF9, accessed on 23.01.2024

A new global order needs to invest in these areas by promoting trade and commerce activities, enhancing the socio-economic infrastructure of schools, colleges, and hospitals, promoting tourism, and building roads and other infrastructure is the viable solution to the existing issues. Businesses cannot operate in isolation, and investment in Corporate Social Responsibility (CSR) reflects their desire to make a meaningful contribution to the socio-economic development of the Maghreb region and their long-term commitment to promote the same in North Africa under multilateralism. The CSR activities of global companies in this region should span the various areas of focus such as education, women empowerment, and rural community development. These global companies particularly are working in the oil and gas sector in this region should engage with the government in exploration and its exploitation and rather strictly avoid any or no promotion of the Black market of oil and gas. It is a fact that this Black market in this region has a billion USD in illicit trade of cheaper oil and gas as compared to market price and national governments in active engagement with global players can curtail this illicit trade. China, India, the EU, America, Japan, and others, the emerging players of the new global order should focus on the community-centered development approach and engage local people in their activities will minimize the extremist threat to the investors. Chinese experience working with their Chinese labor force in the Maghreb region (Libya, Tunisia, Morocco, Algeria, and others) has posed a direct threat from time to time and raised the security threats to the ongoing projects. China's extra cash flow to the Maghreb region and utilizing their domestic resources without the local people's participation adversely affected the region and creating dependency on China is not the right approach to development. 'Concerns about Beijing inconsistently using its no-interference principle to protect and further advance its interests have grown in recent years. The 2019 pro-democracy Hirak movement in Algeria and Chinese support for Algeria's political regime and evacuation of over 35,000 Chinese workers from Libya in 2011, post-pandemic Chinese interference in Morocco and Tunisia domestic affairs' (Elcano, 2024:net) weakens the Maghreb governance and morale of the international investors. The geopolitical right approach in this region is the Public Private Partnership (PPP) on the model of equity basis in which the public partner will be local, provincial, or national government and national private investors will get the opportunity to work with the international investors. Leading private investors in the new world order receive peaceful and safe passage on the sea and land with the active maritime security of the global governance linking security with trade and commerce has been the new initiative implemented for more than a decade now. The leading economic activities either through the Mediterranean, Atlantic Ocean, Suez Canal, Red Sea, or the Indian Ocean are tightly controlled through the different naval fleets belonging to different countries to secure safe passage as a part of global governance. Not only that, USA AFRICOM, a private armed force is actively engaged in this region for peace and security, controlling illicit weapon movement and human migration and smooth flow of all trade and commerce activities. This multilateral arrangement has a congenial dialogue, building consensus in controlling all illicit activities, clearing misperceptions, and resolving any dispute that arises through bilateral or international arbitration.

Maghreb region's natural resources development needs transparency at national and global levels. The national government in this region should be set up on the basis of democratically elected people and procure transparency in dealing with global oil companies. The government oil revenue and income should become a part of the public budget and work in the interest of mass development. Extractive Industry Transparency Initiative (EITI) started with the 24 member countries in Africa (EITI, 2018: net) sharing domestic resource mobilization from extractives, licensing, beneficial ownership, state-owned enterprises, and commodity trading transparency. Maghreb region should adopt EITI membership and receive transparent business and governance for future sustainable development.

8.2 New Geopolitical Strategy

The European proximity of this region enhances its social, economic, and geographic potential naturally. The agricultural commodities of the Maghreb region such as Olive oil and Palm oil are major consumers in Europe, cotton and groundnut producers, are exported to European companies, cocoa and coffee, mainly have a European market and produce a variety of fruits and vegetables, supply to European market.

This region should develop its agricultural sector and promote its agrarian communities in the world market. The untouched sector of this region should be facilitated by providing adequate training, promoting, and building avenues for domestic and international marketing. This region should transform itself from a single economic structure to multi-economic paraphernalia and being connected Geopolitically, it should translate the strategic economic policies into practice. This new strategy should overcome the economic constraints and marketize reinforcing high supporting capacity, a favorable environment, and stable political surroundings. Poverty alleviation and inclusive growth need comprehensive policies, affirmative action plans, and touching different strata of society, which will ensure an affordable standard of living in this region. The uneven lopsided development of the Maghreb countries needs to be addressed inclusively and invite foreign direct investment catering to their needs and curbing inequality accordingly. Geopolitical social resources, social issues, and social harmony need an inclusive strategy as per the local demands and needs that will surely work for the mutual rural–urban sustainable development in this region. Maghreb region's irresponsible exploitation of natural resources has resulted in wealth inequality, enhancing desertification, no policies of proper distribution of water for agricultural, industrial, and domestic usage, and major ignorance toward developing agricultural sector has created social conflicts in this region. Geopolitically, the new strategy should be initiated in the fields of food security, climate change, rural migration, and managing their resource substantially. Today, all Maghreb countries are a member of the AU and need to develop their regional governance institutions. AU peace process and its integrated response to Africa's security and development challenges will work for the peace and development in this region. The prime concern of Geopolitical strategy in this region is the growing concern of checks on corruption and other evil practices, which the national governments should address without delay. The hub for criminal activities, illicit fishing networks, piracy, trafficking of weapons and drugs, and illicit migration of human resources is the Gray zone in the Maghreb region is highly connected to their traditional practices of corruption and needs people's support-driven government policies. This democracy deficit and growing imbalance between levels of human development need a new Geopolitical strategy through Civil Society organizations, NGOs, Self Help Groups, religious leaders, faith communities, and faith-based organizations, which will work in collaboration with the governmental development policies and their action plans. By activating these non-state actors will build peaceful transparency in the society establish the rule of law and promote good governance in the Maghreb region. It will minimize the role of the Islamic State group, and the influence of extremism in Libya and Tunisia, control the violence in Algeria and Morocco, and utilize political fringes as an opportunity in this region. AU should intervene and influence national politics positively, monitor the regular internal behavior of the political setup, and promote human rights in this region. Youth as a dividend in economic development needs a booster of quality education, congenial employment, and social security in this region failing which, the rise of instability and violence will precede.

The geopolitical strategy focusing on developing entrepreneurship, employment-generating education, appropriate development of technology, and utilizing global opportunities at the local level will provide affirmative results in society.

8.3 Conclusion

The sound and jubilant political system is the prime mover for constitutional development, strong electoral behavior, development of different political centers, and strong rule of law in the Maghreb region. Political accountability is answerable to the people right from the local, provincial, and national levels of government functioning, which will develop regional integration in practical terms in this region. Political hostility with neighborhood countries leads to instability, chaos, and an environment of bitterness in the region. This hostility at domestic levels in the absence of strong opposition political parties will lead to high levels of corruption, military upheavals, and social and economic disparities in society. Maghreb region and extended North Africa have to develop their political system, multiparty organizations, regular elections at local, provincial, and regular levels, respect to electoral results at all levels and a comprehensive constitution will only work for a stable political region. Political stabilization in this region will depend on general policy reform and its practices in improving the economic environment. Regionalism will contribute strongly to social and economic development by increasing opportunities to achieve economies of scale, diversifying economic production, improving inter and intra-regional and external trade, and improving policies to strengthen competitiveness. The stable political economy will promote the region's bilateral, regional, and continental relationship and alignment will boost their individual economy and attract global trade in the new world order. Further, the alignments will get affirmative direction, and the preferential trade agreements will lead to free trade agreements in the Maghreb region and will strengthen the global political and economic environment. This environment evolves performance-based structural transformation, utilizing domestic economic profiles and exploring global economic potential in the Maghreb region. Significant changes have been witnessed in the political systems of Egypt, reforms in Tunisia and Algeria, and decentralization in the Moroccan constitution ultimately attracting new international investment in different economic sectors. It indicates the geostrategic location, shape, size, topography, arable land, direct access to the sea, and natural resources of Maghreb countries evolving the investment trends and attracting the new world order for their social and economic development. The Mediterranean linkage between Europe and Maghreb Africa is a source of multi-faceted opportunities and the regional stability of this region will shape trade, commerce, and investment. EU support for the Maghreb as its natural ally will enhance this partnership in the new world order and perpetuate the Geopolitical interests of each other.

References

Antonio, M. (2016). Toward an EU global strategy: Consulting the experts. European Union Institute of Security Analysis. ISBN 978-92-9198-492-3 Available via https://www.iss.europa.eu/sites/default/files/EUISSFiles/EUGS_Expert_Opinions.pdf. Accessed September 21, 2017.

EIA Report. (2018, 8 February). Available via https://www.eia.gov/todayinenergy/detail.php?id=34872. Accessed January 23, 2024.

EITI. (2018, September). EITI in Africa Available viahttps://eiti.org/documents/eiti-africa. Accessed January 20, 2024.

Elcano. (2024). China in the Maghreb: Engagement, perceptions and prospects Available via https://www.realinstitutoelcano.org/en/analyses/china-in-the-maghreb-engagement-perceptions-and-prospects/. Accessed January 24, 2024.

Salsecci, G. (2010). The South Mediterranean Countries: Growth and business opportunities within the framework of relations with the European Union. Intesa Sanpaolo. European University Institute, Florence Robert Schuman Centre for Advanced Studies Global Governance Programme Available via http://cadmus.eui.eu/bitstream/handle/1814/22275/RSCAS_2012_23final.pdf on 20/04/2017 Available via http://www.group.intesasanpaolo.com/portalIsir0/isInvestor/PDF_studi_eng/eos005_eng.pdf. Accessed April 26, 2017.

Suresh, K. (2024). The author was the delegate of World Policy Conference held in Abu Dhabi, 9–12 December 2022 and widely travelled in Maghreb countries.

Westpoint. (2024). Available via https://ctc.westpoint.edu/how-realistic-is-libya-as-an-islamic-state-fallback/. Accessed January 24, 2024.

Suresh Kumar is a Professor in the Department of African Studies, at the University of Delhi. He shouldered the responsibilities as the head and Coordinator of the Department for 2015–19. He is the Chief Editor of the Africaindia.org (http://www.africaindia.org). Professor Kumar is a passionate traveler who has visited forty countries for research and teaching assignments. He articulated his opinion on different aspects of Africa before the Ronald H. Brown Institute (2008), World Policy Council (WPC, 2014), European Commission of African Studies (ECAS), Council for the Development of Social Science Research in Africa (CODESRIA, 2015), IBSA, BRICS-IKS (2016, 2018 and 2019) and many others. He has been visiting since 2012 in Rabat, Casablanca, Marrakesh, and Tantan, the southwestern part of Morocco. He has extensively visited Sahara province and met different tribal groups of Sahara in the year 2018.

Chapter 9
Geopolitical Discourse on the Consequences of Slavery in Contemporary Sudan

Patricia Teixeira Santos and Suresh Kumar

Abstract Slavery in Sudan traces the historical elements of songs, stories, and proverbs of the Nile River, carrying rich historical experiences of different political periods and explaining the livelihood of the Nile society in Sudan. The free-flowing history of the Nile River society, which has experienced the hot golden desert sands and its livelihood, knows the significance of agricultural land and reacted sharply against the beginning of colonialism in 1880. Religion has played an important role in the social awakening, building an approach against slavery and other evils in Sudan. The consequences of slavery and the division of society have geopolitically impacted contemporary Sudan. Ultimately, geopolitical discourse needs to understand the division between Sudan and South Sudan.

Keywords Religion · Colonialism · Missions · Gender · Territories

9.1 Introduction

The purpose of this article is to analyze the relationships established between the various peoples in Sudan and the consequences of slavery in the region, from the existence of the former Mahdist State in Sudan, which originated from the greatest anti-colonial revolt of the nineteenth century and the hierarchies and separations that were structured after the establishment of the Anglo-Egyptian Condominium in 1898.

This article was originally presented in Portuguese at the online international conference: "Sociedades Africanas: Religiosidades, Identidades e ConexõesGlobais", November 2021

P. T. Santos
Department of History and Coordinator of Center for Studies On the History of Contemporary Africa (NEHAC), Sao Paulo Federal University (UNIFESP), Sao Paulo, Brazil

S. Kumar (✉)
Department of African Studies and Centre for African Studies, University of Delhi, Delhi, India
e-mail: skafrica2003@gmail.com

M. Mishra et al. (eds.), *Contemporary Geopolitical Alignment in the Global South*,
https://doi.org/10.1007/978-3-031-81977-3_9

Mahdist Sudan extended into the present-day regions of Southern Egypt, Northern and Southern Sudan, Northern Kenya, and parts of Uganda and Ethiopia. In 2021, we had, as a major milestone in African history, the 140th anniversary of the largest anti-colonial revolt of the nineteenth century, which took place in this confluence of areas, and which stood out both in scope and in the number of people involved in the development of this action in different areas of East Africa, the south of the Arabian Peninsula and India.

The name of the revolt comes from the term "Mahdi" which, in the spiritual and religious path of Islam, means the messianic figure who, initially, should prepare Muslims and the communities protected by them for the end of the world.

The *Parousia* (the return of Christ) would be inaugurated with the return of Jesus and the arrival of the Mahdi was considered an important sign of this eschatology. In Islam, the figure of Christ is seen as a being who went through the maternity of Mary, nevertheless, he is considered Muslim, aware of the unicity of God and his prophet, not sharing his divinity, since he is the result of divine intervention on a woman.

Jesus would be an important prophet of God, but he would not share his divinity. And the phenomenon of the emergence of the Mahdis grew in several African and Asian areas in the context of the occupation processes in the colonial exploitation of Africa and Asia, which gave rise to several resistance movements, among which, those led by Mahdis had considerable mobilizing importance.

Our present theme proposes to talk about the Mahdist Revolt in Sudan, and it is an invitation so that we can develop a reflection on African spaces, which are not necessarily related to our Brazilian history, but which are extremely important for understanding the geopolitical processes of African societies in the 20th and 21st centuries.

The *Mahdiyya* or the Mahdist Revolt, as previously mentioned, was the largest anti-colonial movement in Africa in the second half of the nineteenth century, both in duration (17 years) and in size of Western mobilization for its repression and defeat.

The ability of Sudanese peoples to articulate and communicate with other African spaces to gain support and solidarity was aspect that caught the attention of Father Joseph Ohrwalder, who was a prisoner of the Mahdist State and produced, in great quantity and detail, important reports on the dynamics of the *Mahdiyya*, in which his orientalist (Mitchell, 1995) and disqualifying perspective is noted, at the same time, he scrutinized with interest the dynamics of the Mahdi and his followers in the expansion of support networks and the political significance of the State created by this leadership:

According to the priest:

> All discontented individuals, the very poor, slaves, oppressed, persecuted and many still incited by religious fanaticism associated themselves with him (the Mahdi – translator's explanation), adding those hoping for the spoils, for which the dervishes, skillful rioters, made the greatest promises for this life and the life to come. Finally, you have the support of many slave traders. The latter grew in preference among the rebels. From the moment when Gessi Pasha, with his victory over Suleiman, son of Zubeir, had dealt a major blow to the slave trade and had mercilessly strangled every merchant of that kind he had in his hands, as he told me, the survivors saved themselves by fleeing. Astute by nature, they had perfected themselves. For them, the dervishes promised the spoils and the freedom of the slave trade.

> Those who adhered to the Mahdi went further, becoming more fanatical to the point that they were ready to give their lives for him, though not blindly and without good reason. With this, it is understood how this horde could match the imperial troops and even overcome them, so much so that their captains despised the rebels and did not prepare against them
>
> (Ohrwalder, 1998).

In this excerpt, Ohrwalder indicates a broad support for the *Mahdiyya,* and the recognition of the leadership of the Mahdi, by slave traders and several traders linked to the former Viceroyalty of Egypt and the Ottoman Empire, who perceived themselves to have been aggrieved by the series of concessions that the Ottoman sultan and also of the Egyptian viceroy made to be able to participate in the political and commercial networks of England and France. The text evinces the derogatory view of Islam and restricts the trafficking activity to the African regional context as if it were a "moral" evil of the continent; however, it is an interesting passage showing the surprise in the face of the networks built and the extent of the *Mahdiyya*'s influence in Sudan and allied territories.

A series of punitive expeditions, wars, and conflicts plagued the Mahdist State until 1898, when a broad coalition led by England, in which Canada and the United States participated, being the first foray of these countries into a military conflict in Africa, defeated Caliph Abdulahi, Mahdi's successor after his death in 1885, and his allies.

After the defeat and a series of collective executions, "pacifying" expeditions, and the so-called Fashoda incident, in which England and France almost entered into a conflict of greater proportions, the so-called Anglo-Egyptian Condominium was established in the region and indirect administration was adopted, where the general governor of Sudan was the son of the Mahdi and the entire administrative staff came from the former *ansar*, supporters of the *Mahdiyya.*

The political mediators between the heirs of the Mahdist administration and the British were the catholic missionaries of the former Nigritia Missionary Institute (present-dayComboni Missionaries of the Heart of Jesus), previously mentioned, who had been prisoners of the Mahdists and had the role of interpreters and tutors of the Mahdi's sons.

According to Santos:

> However, much more than a "glorified" perspective of the mission, it was observed that the Church appropriated the experience of the Mahdiyya, and the anti-slavery colonies in Egypt, and created a possibility of mediation between the populations of the southern zone and the British. This action can be seen in the development of disciplinary practices and body control (sexual, behavioral and social control of mission catechumens, from the perspective of author Michel Foucault in his work "Discipline and Punish") in the Catholic mission related to the Anglo-Egyptian administration, which began in 1899. The scope of government action, together with the missionary activity of the Church, involved the ancient populations that had participated in the *Mahdiyya.* In addition, as they were part of the British Empire, these populations began to have more and more contact with the different subjects of that empire, which reflected, in their daily lives, their relations with the colonial world.
>
> (Santos, 2013)

Thus, it is possible to perceive the dynamics of the Christian mission in this Sudanese context, how it interacted with Mahdist society, acting as an interpreter of the Western world for Muslims, and how it mediated with the British, being the depository of the *Mahdiyya*'s heritage.

Throughout the first half of the twentieth century until the independence of Sudan in 1956, the division of the territory was into districts, of which the people of the South, entrusted to Christian missionaries and the teaching of the English Language, and the North, based on the indirect administration of Muslim peoples, creates a division, but one that is permeated by the movement of people, projects, and ideas. There is no discernible religious polarization, despite the Western media's insistence on this, especially in the 1990s, during the conflict in Darfur. The big issue is, in fact, the participation and inclusion of social diversity in the direction of the Sudanese government and public policies. The military governments, especially those of the 90 s of the twentieth century, have maintained the continuity of power of the former northern regional elite, which had the role of exercising indirect administration in the colonial period. However, this line of continuity was updated and transformed, with the increase in demands from other social sectors that wanted qualitative changes and greater inclusion, reinforcing the dictatorial facet.

At the end of 2021, we witnessed the sad news of the worsening of the political situation in Sudan, with the military coup that dissolved a broad coalition, built by civil society, which in 2016 had deposed a dictator who had been governing for over 40 years.

Between 2016 and 2019 there was a period of claims, opening of rights, expansion of women's participation in public and political life, respect for religious diversity, and withdrawal of religious authorities from important public positions. However, this former group linked to Bashir, the deposed dictator, returned to power, since, from the end of the colonial mandate, the main Sudanese riches were being controlled by a military elite that led the end of the colonial process through a reformist path.

This military elite, which took turns in power for over 50 years, had its power questioned in 2016, with President Bashir being denounced and arrested for crimes against humanity. A large portion of his supporters in the military have genocide accusations, especially regarding their participation in the conflict in Darfur, in an atrocious and direct way.

For the first time in Sudanese history, a broad committee was formed involving civil society and different social players who demanded changes for Sudan. This military group, resentful of the loss of access to wealth and power, carried out this current coup d'état.

Raising this topic, in this context, gives visibility to the social struggle of Sudanese universities and civil society that suffer heavily from the control of information, internet, and transport. There are several students and university professors persecuted, arrested, and disappeared right now because of their opposition to the regime, as well as the extreme control of women.

Highlighting this situation reflects a profound criticism of NATO and Western policy, which see the conflict in Sudan as a breeding ground for terrorists and radical Islam, without dialogue with the West, in a secular war of Christians against Muslims.

As we deepen our understanding of Sudan's history, we can criticize the construction of this Western vision, consolidated in the twentieth century, of an anti-Christian, anti-Western, anti-modern Sudan.

Below, we present a history of this research and possible dialogues with the contemporary Sudanese context, motivated by the occasion of the 140th anniversary of the creation of the Mahdist State in 2021.

9.2 A History of the Colonial Conquest of Sudan: The Relationship Between Faith, War, and Slavery

The research we carried out on the *Mahdiyya* and the history of Sudan was only possible because we had the support and solidarity of several people for over 25 years of studies. At the beginning of it, in the early 1990s, there were no incentives to study African history, and much of the research on the subject took place within the scope of social movements since very few universities had disciplines linked to African studies, or when they did, it was in the areas of arts or social sciences.

Since the beginning of this research, we have counted on the enormous solidarity of professors from the library of Cândido Mendes University, who believed in the theme and foresaw its importance for the studies of African history. The Fluminense Federal University allowed me the possibility of access, on two occasions, during my master's and doctorate, the missionary archives, which were the main sources of my research. It was through the University of Padova that we could connect with the University of Khartoum, which had the Muslim sources of the *Mahdiyya*. With this, we were able to establish the dialogue between the sources produced by the Mahdists and the missionaries of the former Nigritia Missionary Institute (present-day Comboni Missionaries of the Heart of Jesus), previously mentioned in this article.

To work with and problematize these missionary sources from the nineteenth and twentieth centuries, which tell the history of the Mahdist Revolt, is to highlight again the importance of universities. It was through this dialogue that we gained access to missionary sources, written in Italian, to personal diaries, letters sent to family members, and mission diaries that informed everything that happened within the mission. We could see, in the writings, stances, and statements of social players about the situations experienced.

On the Muslim side, with sources accessed through publications from the University of Khartum, we saw the Mahdi's letters, which were addressed to various people and institutions, and to missionaries who questioned him. In these writings, we find it interesting to point out that, for Muhammad Ahmad, the Arabic spoken by religious women and nuns was similar to that spoken by slaves, and priests had a much greater knowledge of the language.

With that, complex linguistic exchanges begin, and also the realization of difficulties in understanding and the search for alternative ways of constructing meanings, which made this analysis of missionary sources, produced by Muslims, a very rich process for understanding the complexity of social experience in Mahdist Sudan.

Still regarding missionaries from the Nigritia Missionary Institute, they arrived in Sudan still under the auspices and protection of the Austro-Hungarian Empire and reported to the Austrian authorities. With the dissolution of the latter, the missionaries remained in Sudan, not participating in the Italian unification. This situation weakened them and put them in a position of vulnerability in the Mahdist State since they were not representatives of any of the European states that were belligerent or were interesting to the Mahdi and their followers.

When the Muslim revolt occurred, the missionary women sought support to be rescued but were unable to do so. Sudanese Muslims tried to negotiate with them in exchange for prisoners, and at that moment, they realized that no European nation claimed them, because they did not belong to any active national formation in Sudan at the end of the nineteenth century.

It was through the records of these nuns and also of Muslims that it was possible to observe the intense concern of the Mahdist leaders, who did not understand the reason why no group or nation claimed those religious missionaries. In these sources, one can also see the questioning of Muslim leaders about the type of Christianity proclaimed by those missionaries since they claimed to be bringing Jesus as something new, but there was already secular coexistence between Muslims and Christian communities in Ethiopia and Egypt. Furthermore, these traditional African Christians did not have the attitude of Western Christians, which questioned the civility and autonomy of Muslim peoples.

These sources are very rich since they raise questions about the Christian identity of Italian Christians in Sudan, and it was in these essays that the missionaries would record the political and social situation of the time, including this bewilderment.

The Mahdists had different nomenclatures for the notions of color and race, beyond the outsiders' conception, they rejected the way the West, through the priests and sisters, saw them. And this is a form of resistance since it does not necessarily happen only through warlike confrontation but, above all, in everyday actions, relationships, questioning, and conflicts.

There is an extreme situation expounded in the work "Faith, War and Slavery", published in 2013 and which is also in its expanded translation. The Mahdist Muslims from Sudan, linked to the Sufi confraternities from North Africa and the south of the Arabian Peninsula, are in a region where there has already been centuries-old coexistence of Jews and Christians. However, the arrival of Western priests and Catholic missionary nuns was a novelty for the peoples of Sudan, and their condition of virginity amid a great revolt was considered a political problem and a problem of balance in the relationship with the ancient African Christian communities, that needed to be resolved.

The Mahdists then forced the nuns to marry, since their virginity, in the chaotic context they were in, could not be safeguarded and they would then enter under the guardianship of their spouses. This situation was recorded in the diaries of the priests and in the nuns' testimonies, namely, their private diaries. The Mahdi first, as an alternative, sought to arrange politically well-connected and wealthy Muslims, who by marrying these women would be mutually benefited, gaining prestige in society and protecting the nuns from rape and enslavement. With their resistance, the Mahdi realized, according to the diaries of the nuns and priests, that the rejection was due to fear of apostasy from the Catholic faith and also because the candidates for husbands were not white. The imposed solution was for the nuns to marry white Christians from Greater Syria and one of the religious prisoners.

According to Santos:

> According to Grigolini (one of the nuns who was forced to marry by the Mahdists), the Mahdi himself and his caliphs wanted to marry them, but facing the resistance of the nuns, Muhammad Ahmad (The Mahdi) recommended that they marry white, single men who had been Christians. Thus, according to the nuns and Ohrwalder (one of the prisoner priests), it was believed that a great shock between the spouses would be avoided. Furthermore, those white women would marry men of their own "race
>
> (Santos 2021: 170).

The aspect of distinctions, grounded in the missionaries' racialist views, influenced their assessments of what the Mahdi thought of them and the non-Arab black people who supported him. As a result of these unions between nuns and catechumen women with husbands who supported the Mahdi, the Ondurman*Masalma* emerged. It gained its own status as a community, born from the union of European and African Christian women from the missions with the Mahdists, and their descendants now claim special identity recognition for being part of the modern Sudanese nation and demand recognition of legitimacy and political representation by the Sudanese state.

About the formation of the *Masalma*, Santos tells us:

> The marriage of the missionaries and, in particular, that of Grigolini with Cocorempas, of Corsi and of Ohrwalder, in order to "appease Mahdist pressure", on the other hand, consolidated the leadership of these missionaries in the *Masalma*. The latter was a social subject produced by the governmental action of the Mahdist State or involved by it. As the missionaries' reports had a public/proved dimension, they not only narrated what the *Mahdiyya* was but, from the *Masalma*'s perspective, they performed the mediation role between what the Mahdist historical and religious experience was and their role in these events and in Europe. These dimensions of participants and narrators nourished each other. The marriage of missionaries also represented a possibility, encouraged by the Mahdists, of participating in the construction of the lineage of families that founded the state, a factor that corroborated and deepened the effect of their conversion to the Mahdi and his caliph, hence the personal participation of both at the ceremony's celebration, being public examples of Islamic sexual morality.
>
> (Santos, 2013)

In the context of the *Mahdiyya*, enslavement increased tremendously, for the Egyptians, allies of the Ottoman Empire and the British, subjected a significant contingent of people to wage war against the Mahdists.

During this period, in view of the increase in attacks, the successor of the Mahdi, Caliph Abdulahi, carried out an economic restructuring, using the labor of prisoners, and proposed the opening of work fronts for the cultivation of land and organization of caravans of trade, among other initiatives. The missionaries, who, according to them, had a better Arab than the nuns, were interpreters for the rulers, thus having access to a great amount of information about what was happening beyond the lands controlled by the Caliph.

In the sources, namely, the Mahdi's letters, published by the University of Khartoum, and the work Bayt Al Mal (the treasure book of the Mahdist State) translated by the team led by Anders Bjorkelo Shouk, from the University of Bergen and Sira Al Mahdi, translated and published by the University of Khartoum, we had the unique opportunity to work with the great local African production that describes the missionaries. In the letters written by Muhamad Ahmad, he suggests that the Italian missionaries were not similar to the white merchants and Christians who related to the Mahdi and the *Mahdiyya*, and thus a specific term was created to refer to Italian Christians, calling them Nazarenes.

About these Nazarenes, there are important considerations to be made. The new European nations, which were constituted from 1870 onwards, established compulsory military service, which led to a decrease in the male contingent of missionaries. There was, then, a qualitative change in the missionary contingent, to the point that, during the period of the First World War, most of the Catholic mission was conducted by women. Men, despite becoming a minority, on the other hand, had the most important roles as missionary bishops, heads of mission, and catechists' instructors.

The history of Catholics in the contemporary world is feminine, but we only hear about priests. The priests of the nineteenth and twentieth centuries were the heads of mission stations, apostolic vicars, and administrators, in other words, those who gave guidelines for religious and catechists, who are regional leaders, to lead the mission's life. Twentieth-century Catholic missions were fundamentally carried out by women. This survey data is surprising and makes us review much of this discussion.

At that time, there was also a qualitative decline in the educational background of missionaries. Previously, the missionaries would come from a family of decadent nobility or bourgeois family, they were people with access to good literacy, connoisseurs of Latin, and Greek, and even knew how to handle this ethnocentric and (dis)qualifying repertoire of ideas about African societies.

Missionary reports began to be increasingly concise and written by men who were mission administrators, but who were not directly involved in the mission's daily life. In this sense, the nuns' texts become very precious sources of the reports they made to these administrators, a good part of what the latter reported as the information received from the nuns and catechists. Many times, the missionary priest would merely organize the information received and sign it.

When studying the history of the Mahdiya, we realized this great challenge since the missionary texts had some erudition, such as the ones from Father Ohrwalder, head of the mission station and who became a translator for the Mahdi. However, the great descriptions of daily life and the conditions of the population were made by the nuns.

It is noticed that the priests wrote general texts within a narrative formula, which is that of a colonial world becoming increasingly "barbaric" to justify the presence of missionaries. The sisters, on the other hand, speak of survival, of the support of women, of slaves, they speak of the need for access to these women and of concern for the basic needs of potential catechumen women.

They reported the universe of female everyday life and catechetical work. The missionary reference was often that of the catechist who led the prayers, together with the sisters, and the catechist who organized the reading of the Bible on Sundays, who had permission from the priests administering the mission, together with the mothers, to distribute the host.

For a good part of Sudanese, Ugandan, and Kenyan Catholics, the Christian life was spent living with catechists and nuns. It is possible to notice this through texts that, despite considering their fragile erudition, from a historical point of view are exquisite testimonies of the tensions and conflicts of the African missionary universe. Thanks to these nuns, in the case of the study on the *Mahdiyya* we were able to cross several texts from various origins that helped me to perceive different levels of understanding of analysis, of ethnocentric attitudes, and also of description of mission societies.

In the aftermath of the Second World War Comboni missionaries set out to conduct a major re-reading of the history of ancestral missionary women in the Mahdist State. From the nuns' experiences, and from others in different African spaces, they brought, in their writings, the exposure, in different historical contexts, of gender inequality within the missionary area.

The next item will emphasize the importance of investment in research on African studies and the interfaces that can stimulate fundamental reflections on the Brazilian contribution to African spaces that still need to be known by our historiography.

The history of Mahdist Sudan is vital for understanding several other anti-colonial movements. It needs, as well as other important African spaces, the support and infrastructure necessary for research development in which we can make important contributions with references built in the Brazilian academic field.

9.3 Importance of University Investment in African Studies: The Example of This Research.

It is relevant to stress the importance of Brazilian universities and postgraduate courses in training people who can make coherent research policies and bring to light histories that are different from the ones previously known, and which can bring other dimensions to studies of colonialism, missionary sources, and hierarchical processes of colonial societies. Sudan, for example, was the only colonial formation that had a condominium status. It was not an area of influence, protectorate, or colony, which reveals the interest of several powers over this territory, namely, England, Egypt, and France.

At the time of preparing for my doctorate, the idea was to question the current view of Sudan as an arena for fundamentalists who always hated the West. We sought to bring out the richness, originality, and immense discursive narrative production of these societies that they were in international networks of dialogues and productions of meanings, and the fact that they questioned the newly arrived missionaries and set limits to them. We also crossed a bridge, which appears in the written sources, which is the human dimension of these relationships, especially in the dispute for control of the body, sexuality, and the meaning of life for the five female prisoners.

Revisiting the history that marks the 140 years of the *Mahdiyya* in Sudan is also an invitation to rethink the ways of studying different Muslim contexts and the necessary attention to Islamophobic discourses that produce "convenient enemies," in Peter Gay's sense in his important work, the "The Cultivation of hatred," that further justify the disposal of people and entire societies in the context of profound civilizational crisis that we are going through (Gay, 1995).

Concludingly, studying other African spaces that had no direct correlation with our Atlantic colonial history also allows us to expand theoretical and methodological references. This enables us to revisit old themes, thus bringing new and instigating interpretations, accentuating the richness of African research and universities in the contemporary world. Also, it helps us with issues currently challenging us and that we need to go through globally, such as the impacts of the pandemic on different peripheral social contexts in Latin America and Africa, for instance.

References

Gay P. (1995). The cultivation of hatred. In: *A Experiência Burguesa: da Rainha Vitória a Freud*, vol. 03. (Sao Paulo: Companhia das Letras).

Mitchell T. (1995). Orientalism and exhibitionary order. In *Dirks NB. Colonialism and culture.* Ann Arbor: Michigan University Press.

Ohrwalder, J. (1998). *I mieidieci ani di prigionia: Rivolta e Regno del Mahdi in Sudan* (pp. 45–46). Bologna: EMI.

Santos, P. T. (2013). *Fé, Guerra e Escravidão: uma história da conquista colonial do Sudão (1881–1898)* (p. 203–204, 222 and 286). Editora UNIFESP.

Santos, P. T., Kumar, S. (2021). *Faith war and slavery: A history of colonial conquest of Sudan (1881–1898).* Shivalik Prakashan

Shouk, A., & Bjorkelo A. (1996). The public treasury of the Muslims. Monthly Budgets of the Mahdist State in the Sudan. Bayt Al Mal (1897). Leiden: Brill.

Patricia Teixeira Santos is presently a professor of the Department of History and coordinator of NEHAC at Sao Paulo Federal University (UNIFESP), Brazil. She is also a Collaborating Researcher at the Transdisciplinary Center for Culture, Space and Memory(CITCEM) at the University of Porto (UP), Portugal, and Instituto Les Afriques dans le Monde (LAM)University of Bordeaux- France. She has experience in the area of History, working mainly on the following themes: Africa - historiography, Contemporary Africa- Anti-colonial resistance, Religion and Contemporary African Societies. Since 2022, she has been a guest advisor for the Afro-Brazilian Pastoral of the National Conference of Bishops of Brazil. In 2023, she became an accredited professor on the permanent staff of the postgraduate program in history at the Federal University of Amazonas. Three books, Chapters in different books, and about 60 articles have been published in journals and newspapers. http://lattes.cnpq.br/5205163871513241.

Suresh Kumar is a Professor in the Department of African Studies at the University of Delhi. He shouldered the responsibilities as the head and Coordinator of the Department for 2015–19. He is the Chief Editor of the Africaindia.org (http://www.africaindia.org). Professor Kumar is a passionate traveler who has visited forty countries for research and teaching assignments. He is the Chief Collaborator of the Indian Council of Social Science Research-National Institute for the Humanities and Social Sciences (ICSSR-NIHSS) International Joint Research Project on Religion, Yoga, and Education in India and South Africa. He articulated his opinion on different aspects of Africa before the Ronald H. Brown Institute (2008), World Policy Council (WPC, 2014), European Commission of African Studies (ECAS), Council for the Development of Social Science Research in Africa (CODESRIA, 2015), IBSA, BRICS-IKS (2016, 2018 and 2019) and many others. Eight books, 13 Chapters in different books and about 165 articles have been published in different journals and newspapers. He traveled widely to more than 40 countries in the world for research and teaching assignments.

Chapter 10
Spatio-Temporal Perspectives of Pandemic Risk Reduction: COVID-19 Vaccination in Brazil

Heitor Soares de Farias, Gustavo Mota de Sousa, Vinicius da Silva Seabra, Phillipe Valente Cardoso, and Thiago Jeremias Baptista

Abstract Based on the contributions of Mendez (2023), who understands the pandemic as yet another phenomenon related to the permanent crises (permacrises) of globalized capitalism—hypermobility of people and goods that will be concentrated in urban-metropolitan agglomerations—this chapter aims to analyze the data from the COVID-19 pandemic in 2021, highlighting how countries with a high percentage of the vaccinated population have managed to control the disease, while countries with a low percentage of the immunized population allow the emergence of new strains and new waves of infection. We later expanded the analysis to Brazil, which has been advancing in the immunization of its population despite its president's anti-vaccine stance, and to Rio de Janeiro, which has been emerging in the immunization of its population, with 89% of the population having completed the vaccination cycle, which led to a reduction in the number of people hospitalized and deaths, allowing us to think about reopening and organizing popular festivals, despite the risks. However, when we analyze the Metropolitan region of Rio de Janeiro, which in addition to the capital Rio de Janeiro is made up of another 20 municipalities, the percentage of immunized population is 63%, one of the lowest in the state of Rio de Janeiro. This is worrying, as there are people in the capital who live in other municipalities who have possibly not yet been vaccinated. We need to understand that we are not isolated.

H. S. de Farias (✉)
Department of Geography, Federal Rural University of Rio de Janeiro, Rio de Janeiro, Brazil
e-mail: heisofa@gmail.com

G. M. de Sousa
Department of Cartography and Remote Sensing, Federal Rural University of Rio de Janeiro, Rio de Janeiro, Brazil

V. da Silva Seabra · P. V. Cardoso
Department of Cartography and Geoprocessing, State University of Rio de Janeiro, Rio de Janeiro, Brazil

T. J. Baptista
Department of Human Geography, Federal Rural University of Rio de Janeiro, Rio de Janeiro, Brazil

M. Mishra et al. (eds.), *Contemporary Geopolitical Alignment in the Global South*,
https://doi.org/10.1007/978-3-031-81977-3_10

Territories are connected and population flows can cause distortions in health indicators. In order for us to lead a normal life again, we need to expand vaccination in general, requiring proof of vaccination to allow the free movement of people and thus seek to make everyone aware of the need to be immunized, as vaccines save lives.

Keywords Pandemic · Vaccination cycle · Rio-de-Janeiro · Metropoliton Region · COVID-19

10.1 Introduction

Combating the spread of COVID-19 faces a series of difficulties, as its spread is related to the mobility of people in a globalized and extremely unequal world. There are countries that are on the margins of the immunization process and, consequently, become epicenters of new variants, such as the most recent Ômicron, which is bringing control difficulties to other countries, even those whose population is heading for the third dose of the vaccine. With each variant that emerges, there is concern about the immunization capacity of the available vaccines.

Thus, the distribution of vaccines, the immunization capacity of countries, the emergence of new variants, the demographic and social inequalities of each country, and the capacity of the medical-hospital network, among countless other factors, make studies on the advancement of the disease complex. COVID-19 in the world, and creates the need for this approach to be multitemporal, multi-scale, and multidisciplinary. This means that it is necessary to understand the behavior of the disease in its different phases, reflect on its behavior on the most different geographic scales (global, regional, and local), and that it is important to bring together contributions from the most different areas of knowledge.

In terms of its temporal aspect, this chapter takes an approach aimed at understanding a new phase in the spread of COVID-19 in Brazil, which began with the vaccination process on January 17, 2021, and lasts until November 2021. This new phase is marked by a considerable and gradual change in the contamination curve, until an apparent stabilization that lasted until the moment this study was prepared. As of December 2021, with the discovery of the Ômicron variant, more than one million new cases were registered worldwide in a single day, with Europe breaking daily records for new cases (Pinheiro & Garcia 2021). In Brazil, in December 2021, the 10th, the Ministry of Health website went offline after a hacker attack that removed data, and, weeks later, information about COVID-19 remained inconclusive (Andrade, 2021).

Regarding the multi-scale aspect, we will try to outline different scenarios of the disease, analyzing inequalities in the vaccination process, the impacts of vaccination itself, the number of deaths and the number of infected people in some countries around the world, on different continents at the end of the year 2021. Next, we will make a brief analysis of the situation in the different regions of Brazil, highlighting the same aspects mentioned previously, to finally draw a picture of the pandemic phase for the state of Rio de Janeiro, which together with São Paulo is the main entry points for international tourists in Brazil.

In relation to the need to bring perspectives from multiple areas of knowledge to combat COVID-19, we highlight the contributions of Geography, which in essence is one of the sciences that best responds to the analysis of the distribution of spatial phenomena. Cardoso et al. (2020) highlight that Geography, and its different areas of knowledge, can contribute to the organization and reading of a series of socio-spatial aspects that can support decision-making, ranging from adopting strategies to protect vulnerable and underprivileged social groups, to in the construction of post-health crisis scenarios. Each area of geographic knowledge, with its specificity, presents a valuable range of possibilities for reading the problem. And this is no different when we use the tools made available by Geoprocessing.

Ibañez et al. (2022) state that in the initial moments of the pandemic there was a mistaken idea that the disease would have a "democratic" aspect, in the sense that it would indiscriminately affect all social classes and sectors of society with the same intensity. The authors analyze that, in fact, initially, COVID-19 circulates in Brazil in social classes with greater purchasing power, mainly because the first infected people came from countries where the virus was already circulating. However, what is later noticed is a greater lethality of the disease when it reaches the more popular classes.

Pereira et al. (2021) concluded, based on an analysis of spatial autocorrelation between COVID-19 mortality rates and socioeconomic variables in Brazilian municipalities, generated through the Moran Global Index, that the areas with lower income indicators are, predominantly, those with the highest number of deaths. On the other hand, areas with high purchasing power are generally those with the lowest number of deaths in Brazil. The authors highlight that this can be better perceived when we compare the North-Northeast region, where we find lower income values among municipalities and higher deaths per inhabitants, with the Center-South region, where we have the opposite situation, that is, lower numbers of deaths per inhabitants and municipalities with higher income values.

Another geographical aspect considered important for understanding the spread of the disease, and even for its control, is the analysis based on urban hierarchization, mainly at the regional level. This is because this hierarchy conditions other important factors that are fundamental to controlling the pandemic. One of these factors is the permanent movement of people in search of goods and services, and therefore, the movement of the virus mainly along traffic routes, be they highways, waterways… (Monteiro et al., 2020).

Along with the urban hierarchy, it is necessary to know the infrastructure available for patient care, which is spatially organized according to the level of complexity of the care. A fact observed by Faria et al. (2021), who mapped what we could call the "regional division of care," where some municipalities concentrate on the treatment of serious cases because they have ICU beds, ventilators, a number of doctors, and other fundamental factors related to patient care.

In view of the above, it should be noted that this chapter will therefore undertake to carry out a geographical analysis of the impact of vaccination policies on the distribution of COVID-19 cases, and deaths caused by the disease, at different scales of analysis, with the aim of contributing even more toward the adoption of mitigation measures, and also, to contribute to the construction of scientific records of different phases of the pandemic. To this end, after these introductory considerations, the first section highlights the theoretical framework that allows us to consider COVID-19 in the context of permacrises during the passage through the first decades of the twenty-first century. While the second section addresses the situation of COVID-19 in the world and the third section covers COVID-19 in Brazil. Furthermore, in the fourth section, our analyses present a specific look at COVID-19 in the state of Rio de Janeiro, with these analyses being followed by the final considerations of the article in question.

10.2 COVID-19 in the Context of Permacrises During the First Decades of the Twenty-First Century

The global political space has undergone several transformations in recent decades. Despite the multiplication of regional conflicts in a multipolar geopolitical order, we have seen greater cooperation between States in the face of the different health crises that occurred in the first decades of the twenty-first century. In fact, given the declarations of Public Health Emergency of International Importance, which represent the highest level of alert from the World Health Organization (WHO), we observe that, despite the dissemination of false news and management that underestimated the seriousness of the situation caused by the new coronavirus (SARS-CoV-2), the international system was also marked by assertive geopolitical actions to combat the COVID-19 pandemic.

As we lean into the debate about the COVID-19 pandemic, we understand that a theoretical framework is necessary so that we can reach the empirical aspects related to its multi-scale dimension. Bear in mind that the subsequent sections address the situation of COVID-19 in the world and in Brazil, as well as presenting a specific look at the aforementioned disease in the state of Rio de Janeiro.

In this sense, we start from the perspective offered by Ricardo Méndez who, in his book entitled "Times Critiques for Global Capitalism: A Geoeconomic Perspective," reveals that we can consider the first quarter of decades of the twenty-first century as a critical period. Furthermore, the Spanish geographer argues that this period was subjected to great tensions that significantly affected daily life both collectively and personally, given that the multi-scale aspect of the transformations that occurred were not restricted to the global economic system, but reached several territories and impacted different social classes differently.

From the perspective offered by Méndez (2023), this period was marked by profound crises whose diverse origin, rapid diffusion and global reach resulted in a social malaise that permeated the collective consciousness in recent years, giving rise to the concept of "permacrises," which consists of a period of permanent and successive crises, so that when one crisis ends, another approaches us and, consequently, comes to take its place. In fact, in light of the perspective offered by the referenced author, it is possible to identify, therefore, the occurrence of environmental, financial, health, and geopolitical crises during the first quarter of decades of the twenty-first century.

Regarding these crises, Méndez (2023) points out that:

> On the one hand, they have all been classified as systemic because, whatever their origin, they have achieved a multidimensional character, with direct and interrelated economic, social and political impacts, in addition to transforming pre-existing maps due to their unequal intensity across territories, which also has shown distinct recovery capacity, now linked to the concept of resilience. On the other hand, all of them are threats to our security and well-being that show a global dimension, overcoming all types of border barriers, even with different local manifestations, which highlights the density and speed of current interconnections in a world that works, in good measure, as an open, integrated and hierarchical system, although lacking governance in accordance with this reality.
>
> (Méndez, 2023, p. 9).

Furthermore, the referenced author points out that these recent crises are structural, endogenous, and systemic. And, in this sense, he points out that they are:

> Structural and endogenous because they are not as much the result of a sum of accidents and errors as of the system's own operating logic and, therefore, put into question some of the defining features of global capitalism in recent decades. Systemic because, on the one hand, they affect the global economy as a whole, with interrelated impacts on production, exchanges, consumption and the financing of accumulation processes. But also because it overflows the strict economic scope to focus on social and political relationships, as well as on the behavior of the population.
>
> (Méndez, 2023, p. 18).

Given the above, we understand that there is an environmental crisis that puts the reproduction of the system itself at risk, given that the process of environmental deterioration is related to the disorder in the current functioning of the capitalist system. So, as the aforementioned author points out, the environmental crisis has its best-known manifestation in climate change associated with the accumulation of greenhouse gases and, consequently, the global warming they cause and, consequently, its main consequence is the proliferation of extreme climate events that generate serious impacts on the most exposed territories and populations (Méndez, 2023).

Furthermore, added to the environmental crisis are both the financial crisis that broke out in the first decade of the twenty-first century and the significant ruptures with the previous geoeconomic order represented by the health and geopolitical crises that refer, respectively, to the years between the end of the second and the beginning of the third decade of the twenty-first century. The financial crisis originated within the financial system and materialized with the bursting of the mortgage and real estate bubble, triggered in 2007–2008 with its epicenter in the United States, whose radiation reached different countries around the world that also suffered from high debt. And, consequently, it triggered a major recession whose responses implied the adoption of austerity policies, increased unemployment, and deterioration in the living conditions of the most vulnerable social groups and territories (Méndez, 2023).

Like other epidemics that have occurred in recent decades, the health crisis related to COVID-19 results from a zoonosis whose pathogen resides in an animal species that, at a given moment, comes into contact with human groups. In effect, the emergence of natural environments to serve different human activities implied the possibility of contagion. Whereas, in the case of COVID-19, the health crisis refers to the outbreak of an infectious disease in the industrial city of Wuhan, with the WHO being communicated by the Chinese authorities in December 2019 (Méndez, 2023).

According to Méndez (2023), COVID-19 was not an isolated event, highly unlikely and strange to trends that had already raised alarms at the WHO but was related to yet another of those successive declarations of Public Health Emergency of International Importance that occurred in the first decades of the twenty-first century. In effect, following statements related to the H1N1 pandemic (2009), the international spread of poliovirus (2014), the Ebola outbreak in West Africa (2014), the Zika virus and the increase in cases of microcephaly and other congenital malformations (2016) and the Ebola outbreak in the Democratic Republic of Congo (2018), the WHO declared that the outbreak of the new coronavirus (SARS-CoV-2), responsible for causing the COVID-19 disease, also constituted a Health Emergency Publication of International Importance, on January 30, 2020.[1]

And, subsequently, the aforementioned Organization characterized COVID-19 as a pandemic, on March 11, 2020,[2] when approximately 118 thousand infections had already been recorded in 114 countries and 4,300 deaths reported to the WHO. In fact, given the situation of hypermobility of people and goods, as well as the concentration of the population in urban-metropolitan agglomerations, which characterizes globalized capitalism, we have seen the rapid expansion of COVID-19 on a global scale.

However, in that context, measures to contain the transmission of the disease were related to restrictions on mobility and social interaction in the midst of global hypermobility (Méndez, 2023). Therefore, despite its origins being outside the functioning of the global economy, the consequences of the health crisis related to COVID-19 did not exempt the economic dynamics as its impacts implied retractions in the flows of capital, people and goods given the disorganization in global supply chains production in the contemporary phase of the globalization process.

The perspective offered by Méndez (2023) allows us to understand that this health crisis also had precedents that proved prone to periodic repetition. In this sense, it points out that:

---

[1] Declarations of Public Health Emergency of International Importance (ESPII) are the highest level of alert from the WHO, as provided for in the International Health Regulations. See Pan American Health Organization electronic portal - "History of the COVID-19 pandemic". Available at: https://www.paho.org/pt/covid19/historico-da-pandemia-covid-19. Accessed on: 05 Aug. 2024.

[2] See Pan American Health Organization electronic portal - "History of the COVID-19 pandemic". Available at: https://www.paho.org/pt/covid19/historico-da-pandemia-covid-19. Accessed on: 05 Aug. 2024.

> Thus, since the beginning of this century, the world has recorded the appearance of new epidemic diseases that affect specific territories. This was the case of SARS that emerged in China (2002), the H5N1 or avian flu, whose initial outbreak appeared in the same country (2003), the H1N1 or porcine flu, which was initially detected in Mexico (2009), the MERS of Middle East (2012), Ebola in Central and Western Africa (2014), Zika that emerged in Latin America (2015). All this is linked to the persistence of re-emerging diseases, which appear periodically or are endemic in certain regions, such as malaria, cholera, tuberculosis, dengue, yellow fever, chikungunya, Nile and Nipah fever, AIDS (Méndez, 2023, p. 27).

That said, it is worth noting that, while we were still experiencing the immediate collapse in the circulation of people and goods due to confinement measures, as well as a period of economic instability and significant numbers of reported cases related to the disease, as well as significant numbers of deaths from COVID-19 reported to the WHO,[3] there was a geopolitical crisis related to Russia's invasion of Ukraine in 2022. In fact, this geopolitical crisis occurred even before the WHO declared the end of the Public Health Emergency of International Concern regarding COVID-19[4]. And, given the above, it is understood that the succession of challenges to the international system persists and the concept of "permacrises", as proposed by Méndez (2023), seems increasingly pertinent.

The geopolitical crisis, whose impact was immediate, is related to the invasion of Ukraine by Russian troops, in February 2022. Given that, the conflict in the Eastern European region, there was an increase in tensions between the great powers, the reactivation of political-military alliances such as, for example, NATO to face the threat from the eastern portion, as well as the approval of financial and commercial sanctions against Russia. In effect, we saw the emergence of difficulties in reactivating the post-pandemic economy as there was an increase in inflation and a reduction in economic growth prospects (Méndez, 2023). And, in the meantime, it is clear that the international system is going through a period of "permacrises". In this sense, it is worth noting that even in the face of the geopolitical crisis related to Russian-Ukrainian tensions in Eastern Europe, the number of cases and deaths from COVID-19 reported to the WHO continues, whose numbers reported by this Organization represented, respectively, around 53,900 cases and approximately 789 deaths, as of July 28, 2024[5].

[3] See the World Health Organization (WHO) electronic portal. Available at: https://data.who.int/dashboards. Accessed on: 11 Aug. 2024.

[4] See Pan American Health Organization electronic portal - "History of the COVID-19 pandemic". Available at: https://www.paho.org/pt/noticias/5-5-2023-oms-declara-fim-da-emergencia-saude-publica-importancia-internacional-referente. Accessed on: 11 Aug. 2024.

[5] See the World Health Organization (WHO) electronic portal. According to data made available by the WHO—WHO COVID-19 dashboard - it is understood that there were 53,900 cases of COVID-19 reported to the WHO, as well as 789 the number of deaths reported to the WHO on July 28 2024. Available at: https://data.who.int/dashboards. Accessed on: 11 Aug. 2024.

10.3 COVID-19 Situation in the World

According to the World Health Organization's information panel (WHO, 2021), the world had a total of 262,866,050 total cases of COVID-19, from the beginning of the pandemic until December 2, 2021, distributed across all countries. The number of deaths had already reached a total of 5,224,519 people, with 7,885,790,810 doses of vaccine already administered. It is worth noting that these numbers reflect notifications made by countries in a process in which delays in notifications are common.

The distribution of total confirmed cases by WHO regions, recorded since the beginning of the pandemic until December 2, 2021, showed the Americas with the highest values, above 36%, followed by Europe with 33.37% (Fig. 10.1). However, with Ômicron, Europe showed a worrying growth in the number of cases, with four consecutive weeks of increases of around 10%. Meanwhile, in the Americas, what was observed were fluctuations with trends toward a reduction in the number of weekly cases, without yet registering the impact of the new strain.

In Southeast Asia, the figures were around 17% of the cases already registered, with a strong tendency toward a reduction in the number of cases, which in that period already registered drops of more than 10%. The other WHO regions are subdivided into the Eastern Mediterranean, which registered 6.4% of the cases already recorded, in addition to the Western Pacific, with 3.9% of cases and Africa, with just 2.4% of all records.

On another scale of analysis, 7 countries in Europe and America, were chosen to understand the behavior of the disease and the impacts of vaccination at this level of analysis. The choice was based on the analysis of different profiles that helped verify the change in the pandemic in different situations. To this end, Brazil, the United States of America, Chile, Argentina, Portugal, Germany, and the Russian Federation were analyzed.

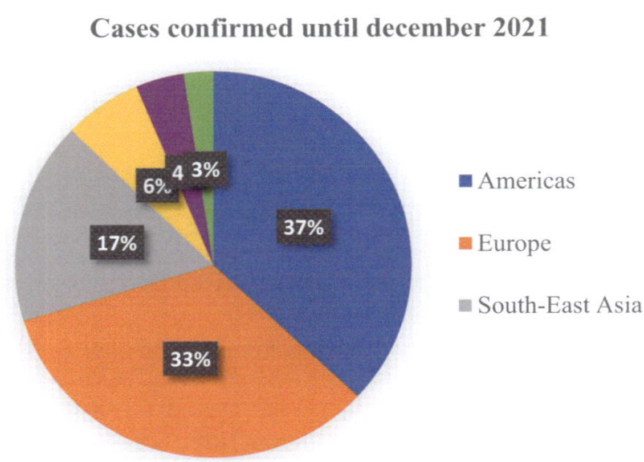

Fig. 10.1 COVID-19 cases confirmed until December 2021 (Prepared by the authors from the datasource of WHO, 2021)

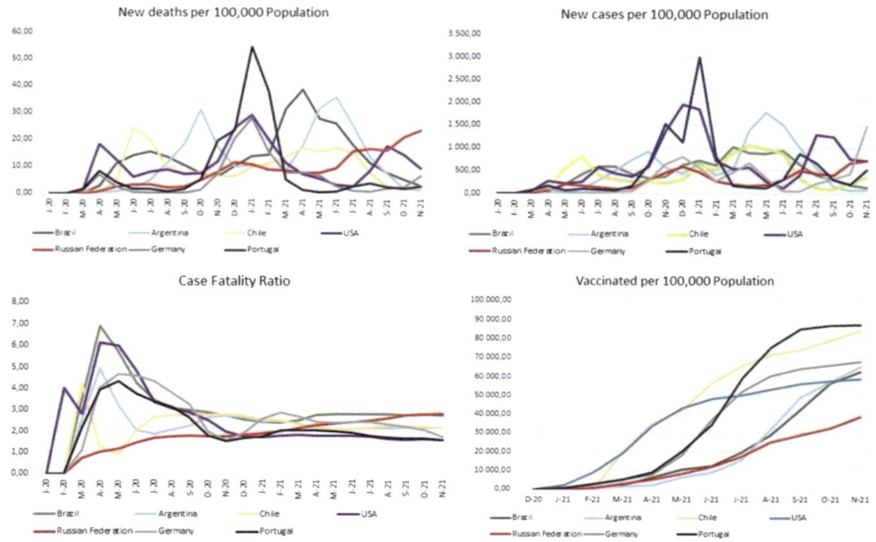

Fig. 10.2 New deaths, cases and vaccinated per 100,000 population, and case fatality ratio in some countries of the world (Prepared by the authors from the datasource of WHO, 2021)

In the first stage of the investigation, the following parameters were analyzed: New deaths per 100,000 population; new cases per 100,000 population, vaccinated per 100,000 population, and case fatality ratio for all countries analyzed. It is important to highlight that all data were normalized by the estimated population of the countries. Secondly, we analyzed the impact of vaccination on reducing new deaths, using the graphs generated for each of the seven countries (Fig. 10.2).

The first important aspect that we can observe, based on the graphs generated with Our World In Data data (Ritchie et al., 2021), is the change in fatality ratio values from the first months of the pandemic to the following moments, noticed in all countries analyzed. It can also be seen that in these first months, the numbers are higher in the USA and Brazil, which gradually decrease until they are close to the values in other countries.

When we compare the graph of monthly deaths, or new deaths per 100,000 population, with the graphs of new cases per 100,000 population, we see different situations between countries. Even though the graphs are bimodal for most countries, which suggests the occurrence of a first and second wave, these occurrences have different temporalities. The second waves in Brazil, Argentina, and USA were more impactful than the first waves. In Chile, the first wave was more acute than the second. In the midst of this analysis, we draw attention to the Portuguese case, which presents a very controlled first wave, at the same levels that were observed for Germany, but a more aggressive second wave among the countries analyzed.

When looking at the graph of vaccination per 100,000 population, we also see that vaccination is on a growth curve in all countries, which clearly shows inequality between them. While countries like Chile and Portugal have reached numbers close to 80% of the population vaccinated, and others, like Brazil and Argentina, continue on a very upward vaccination curve, the Russian Federation shows a situation with less than half of the population vaccinated and a curve that is still very timid of immunization.

The graph that articulates new deaths per 100,000 population and the percentage of vaccinated among the countries analyzed (Fig. 10.3) shows that vaccination ushers in a new phase of the pandemic. After the start of vaccination, a progressive reduction in the number of new deaths was observed. The only exception is the Russian Federation, which is still experiencing a significant increase in new deaths, which can be explained by the low percentage (below 40%) of the vaccinated population.

10.4 COVID-19 in Brazil

According to estimates by the Brazilian Institute of Geography and Statistics (IBGE), Brazil reached a population of 213.3 million inhabitants in 2021, with a concentration of the population in large cities (IBGE, 2021a, 2021b). There are 17 municipalities with more than 1 million inhabitants (14 of them are capitals). This group accounts for 21.9% of the population or 46.7 million people. Another 49 have more than 500 thousand residents and 326 have more than 100 thousand people.

It is also important to highlight that Brazil is organized politically and administratively into 5,570 municipalities that are grouped into 27 federative units, which are grouped together and, therefore, form the North, Northeast, Southeast, South and Central-West regions (Fig. 10.4). However, the population is not distributed equally across these units, as more than 42% of the population is found in the Southeast region of the country, mainly in the states of São Paulo (42 million), Minas Gerais (20 million), and Rio de Janeiro. January (16 million) (IBGE, 2021a, 2021b).

As of December 5, 2021, Brazil had more than 22.14 million confirmed cases of COVID-19, which means that approximately 10,537 for every 100,000 Brazilian inhabitants had already contracted the disease. Of this total, 615,636 people died, which represents a mortality of 293 people per 100 thousand inhabitants (Secretarias Estaduais de Saúde. Brasil, 2021).

When we analyze the distribution of COVID-19 cases by regions of Brazil (Fig. 10.5), we see that the highest concentration of the disease is in the Southeast region, followed by the South and Northeast regions, within a logic that follows that of population distribution in Brazilian territory. However, when we carry out an analysis based on the numbers of the disease, relativized by the number of people living in each of these regions, we have a picture of greater concentration, both of accumulated cases and accumulated deaths since the beginning of the pandemic, in the Midwest region.

Sousa et al. (2022) highlight that in Brazil, differences in basic infrastructure in large centers and outskirts, especially with regard to access to minimum hygiene conditions, safe water, and sewage treatment, cause major differences in terms of

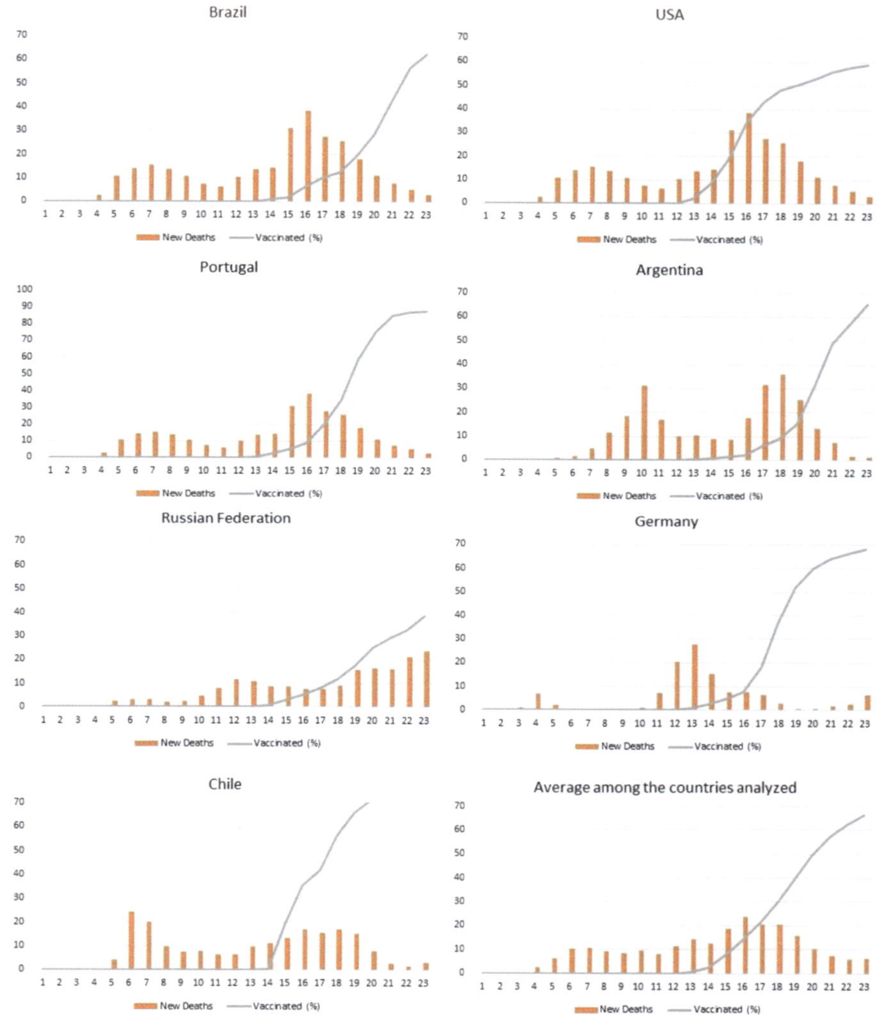

Fig. 10.3 New deaths per 100,000 population and percentage of vaccinated among the countries analyzed (Prepared by the authors from the datasource of Ritchie et al., 2021)

impacts caused by COVID-19. Cardoso et al. (2022) point out that these regional differences increase the spread of the virus, because when beds are not available in certain locations, alternatives are sought in adjacent areas, and when we move patients, we also mobilize the virus.

According to data from Our World In Data (Ritchie et al., 2021), in Brazil, by December 2021, 325 million doses of the vaccine had already been administered, with approximately 141 million people completely immunized, which represents

Fig. 10.4 Political division of Brazil (*Source* Adopted from IBGE, 2021a, 2021b)

approximately 66, 6% of the entire population of the country. This shows that, even though the vaccination process began late, much later than in some other countries, with the first person vaccinated in January 2021, Brazil has a very advanced logistics and service network for vaccination, guaranteed by the Unified Health System (UHS).

However, it is important to highlight that vaccination does not occur equally throughout Brazil. According to the Consortium formed by the press vehicles Folha, UOL, O Estado de S. Paulo, Extra, O Globo, and G1, in December 2021, in the state of São Paulo, 84.5% of the entire population had already been vaccinated with at least one dose, with 77.8% having already received both doses, and 17.2% having already received the third dose, or booster dose. On the other hand, in the same period, in the state of Amapá, only 59.8% of the population had received at least one dose of the vaccine, with only 39.1% receiving both doses and only 1.1% the booster dose. In a more general analysis, we noticed that the states in the Northern region of Brazil have lower percentages of their population vaccinated.

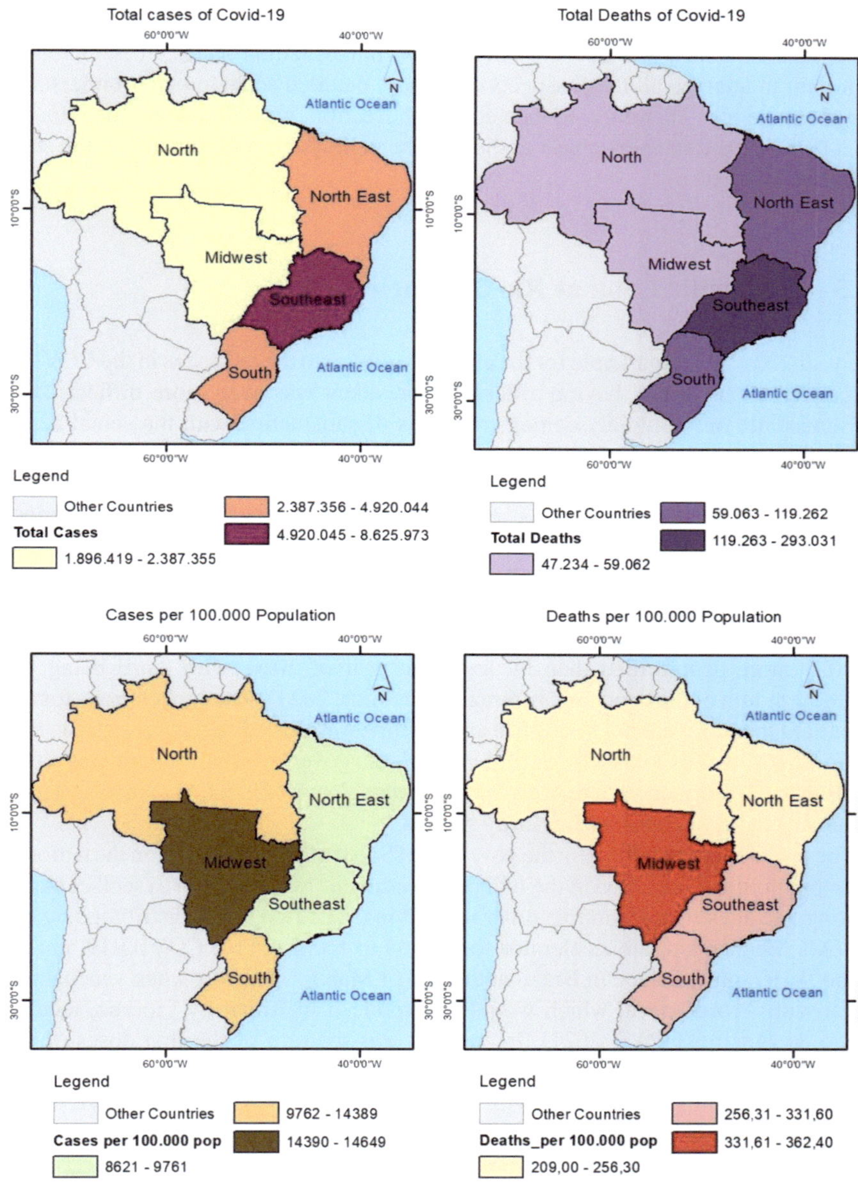

Fig. 10.5 COVID-19 in Brazil (*Source* Prepared by the authors)

The differences in infrastructure, in the numbers of cases and deaths and also in vaccination rates, are also considerable within Brazilian states. It is therefore important to analyze all these aspects on a more detailed scale to better understand this phenomenon. Therefore, in the subsequent section, we will discuss the impacts of COVID-19 and the new phase of the disease in the Metropolitan area of the state of Rio de Janeiro.

10.5 A Specific Look at Rio de Janeiro

The year 2020 was remarkable for the entire world due to the advances in the COVID-19 pandemic. In Brazil, having a denialist president was even more difficult. Jair Bolsonaro not only publicly demonstrated his dissatisfaction with the social isolation strategies published by the WHO, and followed by local managers, but he also disrespected them. He did not wear a mask and promoted motorcycle rides and gatherings. Of all his negative attacks, perhaps the harshest was against the vaccine. Deliberate government actions caused a delay in the start of vaccination and the result was 194,949 deaths from COVID-19 in Brazil in 2020 (Schreiber, 2021).

In 2021, the first months were even more complicated. As of April 24, there had already been 195,848 deaths from COVID-19, that is, in the first four months of 2021 more people died than in the entire year of 2020, with April being the deadliest month of the pandemic in Brazil (Barcellos, 2021). The expectations for the remaining months were not good at all and all hope was placed on the vaccination that began slowly, thanks to the efforts of the São Paulo government, through the Butantã Institute in the production of Coronavac in partnership with China. Vaccination began in January in some states with priority groups.

The prominence achieved by the governor of São Paulo, João Dória, on the national scene put political pressure on the federal government, which began to seek contacts to bring vaccines from countries other than China. The first doses distributed by the COVAX Mechanism, an accelerator for access to tools against COVID-19, co-led by the WHO, only arrived in Brazil at the end of March. That was when vaccination began with AstraZeneca, which would be produced in Brazil by Fiocruz, located in Rio de Janeiro (OPAS, 2021). In April, it was the turn of the first doses of the Pfizer/BioNTech vaccines, produced in Belgium, which are part of the agreement signed by the Ministry of Health with Pfizer, celebrated only in March 2021. Thus, vaccination gained a faster pace from April onwards, and it accelerated in a calendar that progressed according to age groups (BBC News, 2021).

In June 2021, with the population over 80 years old immunized with two doses and those over 60 years old with the first dose, there was a reduction in the number of deaths from COVID-19 in this age group. Given this, there was also an increase in the number of deaths in lower age groups, which, however, was reduced as vaccination progressed (Nitahara, 2021). We ended 2021 with a high percentage of immunization of the population aged twelve and over, with two doses of the vaccine, and with the number of daily deaths falling day after day. The challenge now is to vaccinate children aged between 5 and 11 years old, which, once again, has been met with resistance from Jair Bolsonaro's government.

Despite the government, the easing of the pandemic in Brazil allowed the use of masks in open spaces in the state of Rio de Janeiro to become more flexible (Campos, 2021), while in the city of Rio de Janeiro, the country's cultural capital, face-to-face classes were resumed, as well as allowing the presence of the public in football stadiums and concert halls. They even announced the return of Carnival for 2022, as there are currently no restrictions on parties, according to Rio de Janeiro's own scientific committee (O GLOBO, 2021). There are only 12 people hospitalized with COVID-19 in the city of Rio de Janeiro.

However, since the discovery of the Ômicron variant and with the new outbreak of contamination that has been frightening Europe, with more than 5 thousand flights canceled at the end of 2021 (Mclymore, 2021), preventively, in Brazil, New Year's Eve parties were canceled. In Rio de Janeiro, Carnival, an important popular manifestation of Brazilian culture, became a doubt. Next, we analyze data on the evolution of the pandemic and vaccination in Rio de Janeiro, and government regions (Fig. 10.6), to understand whether this euphoria or concern is justified.

The graph of total monthly hospitalizations and the moving average (Fig. 10.7) shows that there has been a reduction in the number of new cases in the state of Rio de Janeiro and, in the month of November, with 21,840 cases, we reached the lowest numbers since April 2020, when the pandemic begins to gain strength in Brazil.

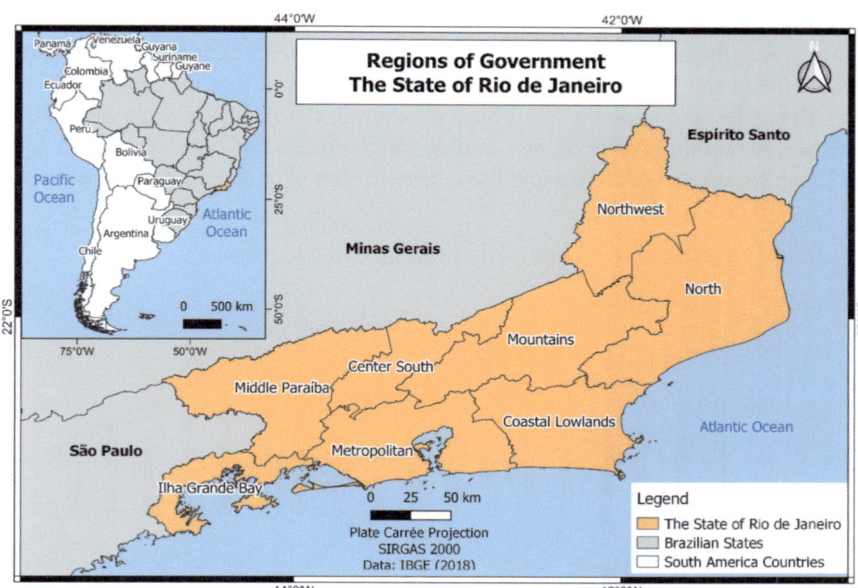

Fig. 10.6 Regions of Government in the State of Rio de Janeiro 2018 (*Source* Adopted from IBGE)

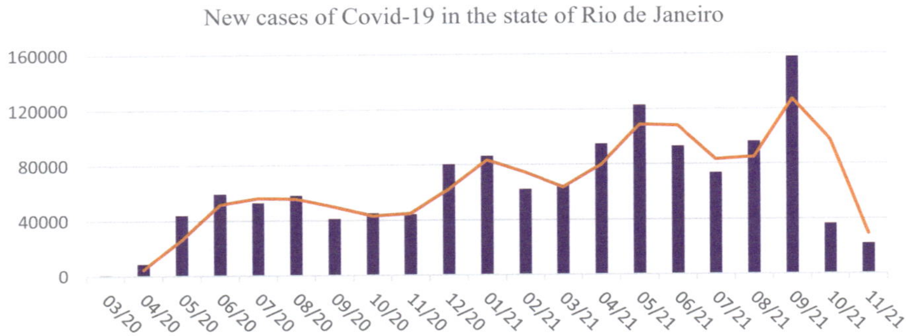

Fig. 10.7 New cases of COVID-19 in the state of Rio de Janeiro. The purple bars are the monthly cases and the orange line is the moving average of cases (*Source* Prepared by the authors)

Following the reduction in the number of registered cases, we also saw a significant reduction in the number of deaths from COVID-19 in the state of Rio de Janeiro after reaching the maximum number of new monthly deaths in April 2021 (Fig. 10.8). After the most lethal month in the series, when more intense vaccination also began, with the arrival of the first doses of Astrazeneca and Pfizer, there was a reduction in deaths in May 2021 and in subsequent months. In November 2021, 704 deaths were recorded, being the lowest value in the series, only higher than the month of March 2020, when the first death from COVID-19 was recorded in the state of Rio de Janeiro.

Analyzing the two previous graphs—numbers of new cases and new deaths from COVID-19 in the state of Rio de Janeiro—it is observed that the number of new cases fluctuates and shows a small increase even in the month of September 2021. However, the monthly total of new deaths, which reached its highest value in April 2021, has since reduced, following the intensification of vaccination. So we can say

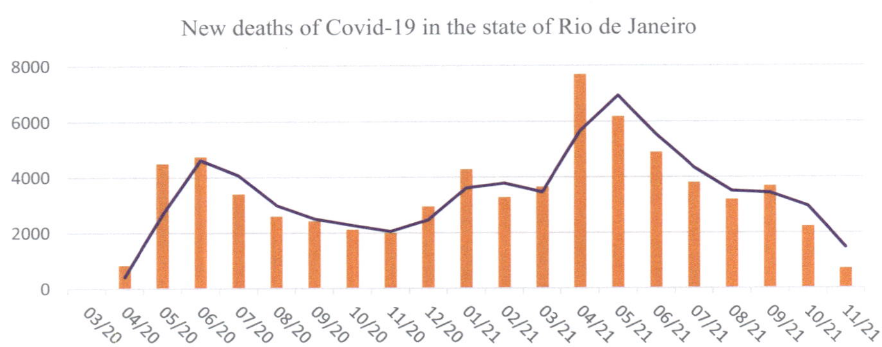

Fig. 10.8 New deaths of COVID-19 in the state of Rio de Janeiro. The orange bars are the monthly deaths and the purple line is the moving average of deaths (*Source* Prepared by the authors)

that the vaccine has proven to be fundamental in controlling deaths, although new cases continue to be registered.

Given this information, we began to seek to understand how COVID-19 spread in the state of Rio de Janeiro, based on its subdivision and grouping of municipalities by government regions. COVID-19 first arrived in the city of Rio de Janeiro and soon reached the entire Metropolitan region which, in addition to a greater population concentration, has an intense flow of people (Lucena et al. 2020). Workers living in the outskirts commute daily, making highways, especially public transport, potential vectors for COVID-19 transmission. Subsequently, the Metropolitan region began to establish contact with other regional centers from its periphery, and from these with the capital, with a large number of cases and a great capacity for spreading COVID-19 to municipalities in the interior of the state (Monteiro et al., 2020).

The following graph shows the Prevalence of COVID-19 among the regions of the state of Rio de Janeiro (Fig. 10.9). Prevalence Rate is the measure of the total number of existing cases of a disease in a period of time and a given population. Prevalence is an indicator of the magnitude of the presence of a disease in the population (Barbosa et al., 2010). The calculation is: Prevalence Rate = (Total cases / Population) × 100,000, where the result presents the number of cases per 100 thousand inhabitants, allowing municipalities with very different populations to be compared.

Throughout the data series, the Metropolitan region had the lowest prevalence for COVID-19, despite having around 61% of COVID-19 cases in the state of Rio de Janeiro, which is a lot. The prevalence rate is low because the Metropolitan region has around 73% of the state's population. Thus, compared to the populations, the prevalence in other regions of the interior of the state was higher, such as in Noroeste Fluminense where 3.5% of cases were registered with less than 2% of the population

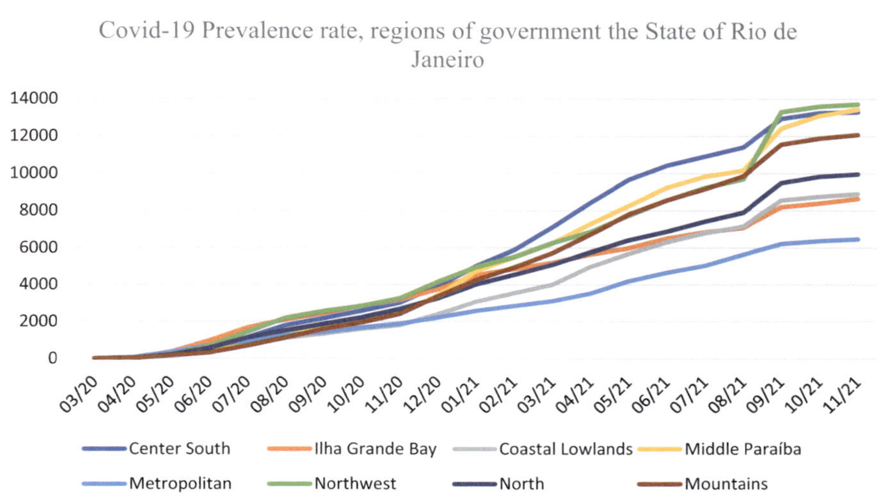

Fig. 10.9 COVID-19 prevalence rate in Rio de Janeiro, by government regions (*Source* Prepared by the authors)

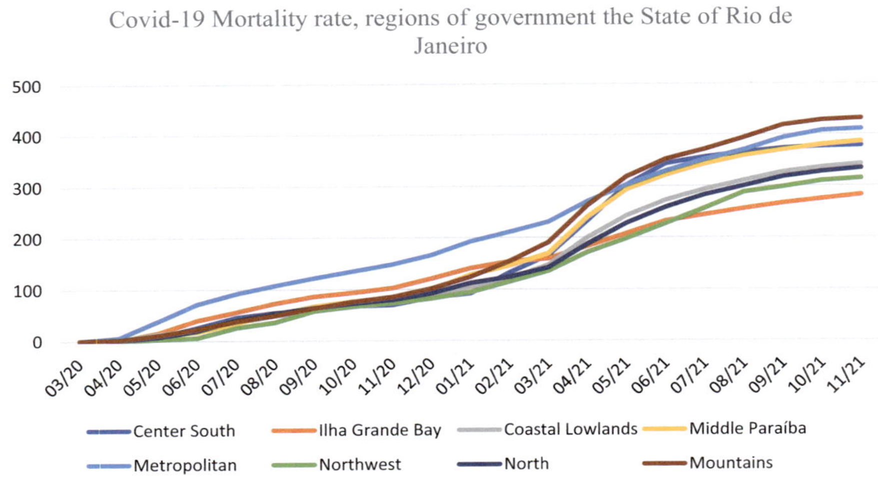

Fig. 10.10 Mortality rate from COVID-19 in Rio de Janeiro, by government regions (*Source* Prepared by the authors)

of the state of Rio de Janeiro. This may be a consequence of the insufficient number of tests available or applied in the Metropolitan region, for example.

To compare deaths, the mortality rate was used, which is a measure of the total number of deaths over a period of time and in a given population. The calculation is: Mortality Rate = (Total deaths / Population) × 100,000, where the result presents the number of deaths per 100 thousand inhabitants, which allows comparing mortality in municipalities with very different populations (Fig. 10.10).

The Mortality Rate graph (Fig. 10.10) shows that the Metropolitan region presented much higher values than other regions during the first year of the pandemic. However, as of May 2021, the Metropolitan region became the second largest in the state of Rio de Janeiro, as it was surpassed by the Serrana region. It is important to note that there was a more intense increase in the initial months of 2021, but that it stabilized after the start of the broader vaccination period.

The increase in mortality in the interior of the state is the result of flows that originate in the Metropolitan region itself, due to the extrapolation of contacts between its periphery and other sub-centers, causing COVID-19 to circulate more intensely in these regions. The Map that shows mortality in Rio de Janeiro (Fig. 10.11) allows us to identify that the regions bordering the Metropolitan area, mainly to the north, which are Middle Paraíba, Center South and Mountains, have mortality rates similar to the Metropolitan region.

When calculating the fatality rate, which is a measure of the total number of deaths in the population that was affected by the disease, it expresses the severity of the situation, indicating the percentage of people who died from this disease (Barbosa et al., 2010). The calculation is: Fatality Rate = (Total cases / Total deaths) × 100, where the result reflects the percentage of deaths among patients (Fig. 10.12).

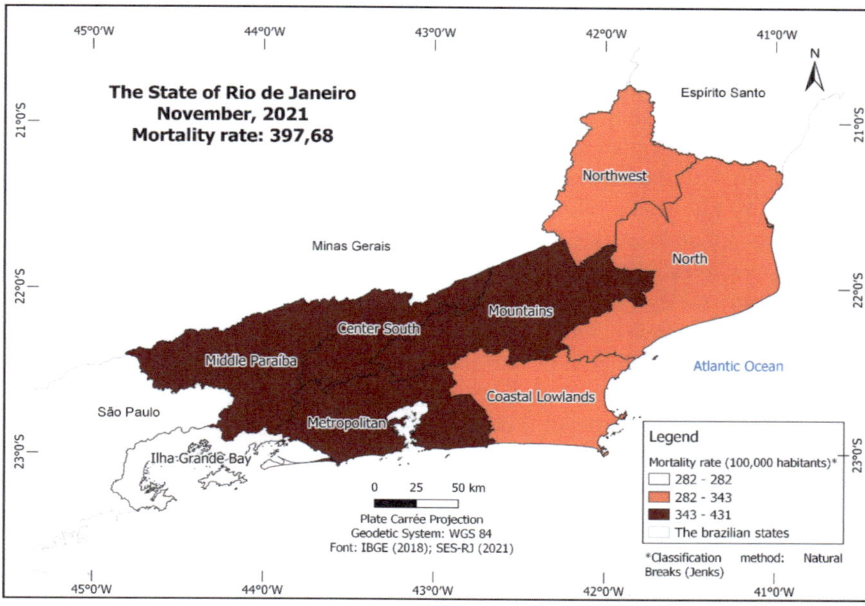

Fig. 10.11 Map of the COVID-19 Mortality Rate in the state of Rio de Janeiro, by government regions (*Source* Prepared by the authors)

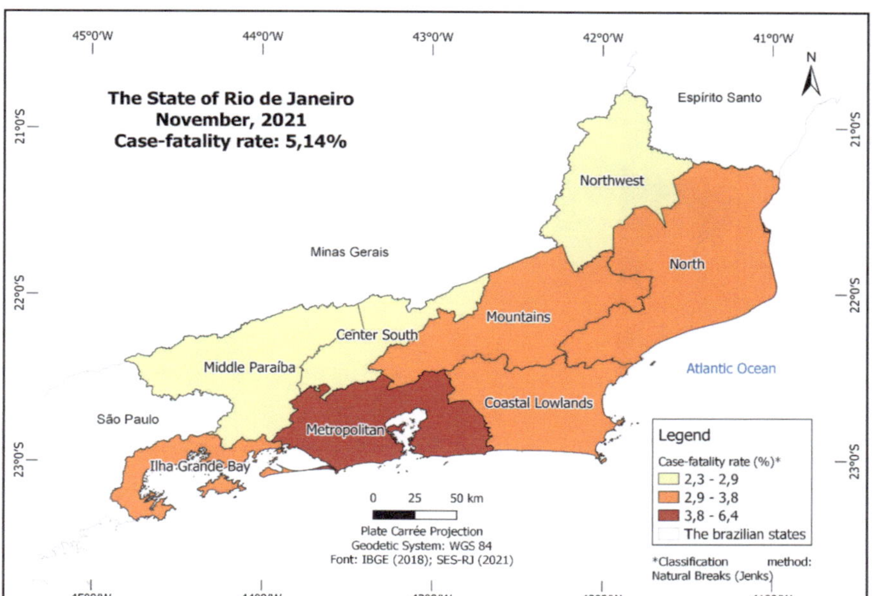

Fig. 10.12 COVID-19 Death Rate Map in the state of Rio de Janeiro, by government regions (*Source* Prepared by the authors)

In the state of Rio de Janeiro, despite many cases in the Metropolitan region, the percentage was low in relation to the total population, and thus the prevalence rate was low. However, as the mortality rate was high, lethality reached high percentages in this area, much higher than in other regions. Therefore, COVID-19 was more serious in the Metropolitan region (6.3%) than in any other region in the state of Rio de Janeiro. The Coastal Lowlands, in second place, recorded a 3.8% fatality rate. The lowest fatality rate was in Noroeste Fluminense, a region furthest from the capital and which is not an interstate highway crossing area.

Finally, we investigated the percentage of the vaccinated population aged 12 and over, which was the age group with vaccines available at the end of 2021 (Fig. 10.13). The regions that showed the highest mortality from COVID-19 were those that achieved a higher percentage of vaccinated population—Middle Paraíba (72.7%), Center South (72.4%) and Mountains (70.4%)—which it may be a reflection of fear of contamination and death. The exception was the Metropolitan region, which had the highest mortality rate, the highest fatality rate and the lowest percentage of vaccinated population (63%), only higher than Ilha Grande Bay (45.5%).

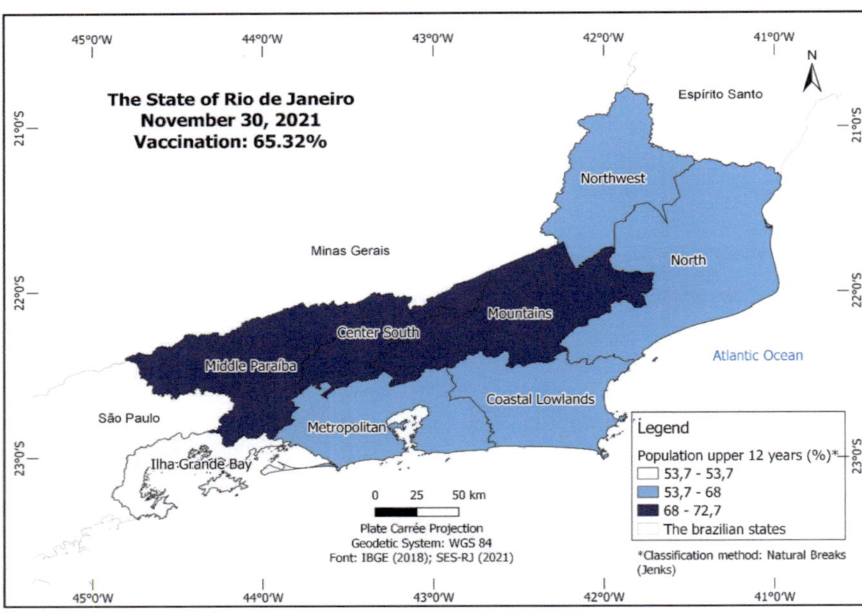

Fig. 10.13 Vaccination map in the state of Rio de Janeiro, by government regions (*Source* Prepared by the authors)

10.6 Final Considerations

The studies proposed here show that COVID-19 has its dispersion related to various socio-spatial aspects, but, above all, being a phenomenon that is related to globalized capitalism in the face of a situation of hypermobility of people and goods, as well as the concentration of the population in urban-metropolitan agglomerations that characterize it. Thus, in fact, it can be studied in light of Geography, which can, as a science, bring a series of contributions to decision-making related to the control and mitigation of the pandemic.

We can also point out that the beginning of the vaccination process inaugurates a new phase of the pandemic, and that this new phase makes studies on different scales even more necessary, since the countries, states and municipalities that are able to vaccinate their population more quickly and efficiency experience a sharp drop in the number of infected people and deaths caused by the disease more quickly. This aspect also makes clear the fact that countries' unequal access to vaccines, as well as other difficulties in this process, make control of the COVID-19 pandemic even more unbalanced.

Another important fact is not to forget to consider the regions to which the territories are located. On a global scale, the networks of flows facilitated by globalization, on a local scale, in commuting movements carried out between municipalities in the same region, mainly metropolitan ones. These population flows can cause significant changes in health indicators, creating false expectations.

Take as an example the municipality of Rio de Janeiro, which reached 89% of its population vaccinated and, as it showed reductions in the number of hospitalizations and deaths, imagined being able to hold popular parties on New Year's Eve and Carnival. However, the municipality of Rio de Janeiro is a metropolis and together with twenty other municipalities form its Metropolitan region. No other of these have such high percentages of the vaccinated population, with two having less than 40% of the vaccinated population.

Thus, even though Rio de Janeiro has a very high percentage of the population immunized, there are people living in other municipalities in the capital who have possibly not yet been vaccinated. There is also the possibility that this floating population, coming from other municipalities, could have accessed the health system in the capital and been vaccinated. In other words, how many of these doses applied were to residents of other municipalities? In this case, there is a lower percentage of vaccination for residents of the city of Rio de Janeiro itself.

Therefore, even though the population of Rio de Janeiro is widely vaccinated, it is necessary to understand that they are not isolated, they do not live in a bubble, and therefore it is necessary to expand vaccination in their region. Caution must be exercised before allowing popular festivals, before allowing large gatherings. Proof of vaccination must continue to be required to allow the free movement of people and seek to make everyone aware of the need to be immunized, as vaccines save lives.

References

Andrade, H. (2021). Site do Ministério da Saúde sofre ataque hacker durante madrugada e sai do ar. CNN Brasil, São Paulo. 10.12.2021. Disponível em: https://www.cnnbrasil.com.br/nacional/site-do-ministerio-da-saude-sofre-ataque-hacker-durante-madrugada-e-sai-do-ar/. Acessado em: 30 de dezembro de 2021 (in Portuguese)

Barbosa, J. et al. (2010). Módulos de princípios de epidemiologia para o controle de enfermidades. Módulo 1: apresentação e marco conceitual. Brasília: Organização Pan-Americana da Saúde, 48 p., 2010 (in Portuguese)

Barcellos, R. (2021). Número de mortes por COVID-19 no Brasil em 2021 já supera todo ano de 2020. CNN Brasil, São Paulo. 25.04.2021. Disponível em: https://www.cnnbrasil.com.br/saude/numero-de-mortes-por-covid-19-no-brasil-em-2021-ja-supera-todo-ano-de-2020/. Acessado em: 26 de dezembro de 2021 (in Portuguese)

BBC News. Vacina da Pfizer: Brasil recebe 1° lote do imunizante com 4 meses de 'atraso'. Brasília. 21.03.2021. Disponível em: https://www.bbc.com/portuguese/brasil-56930584. Acessado em: 26 de dezembro de 2021 (in Portuguese)

Campos, A. C. (2021). Estado do Rio de Janeiro flexibizará uso de máscara em local aberto. Agência Brasil, Rio de Janeiro. 27.10.2021. Disponível em: https://agenciabrasil.ebc.com.br/saude/noticia/2021-10/estado-do-rio-de-janeiro-flexibilizara-uso-de-mascara-em-local-aberto. Acessado 26 de dezembro de 2021 (in Portuguese)

Cardoso, P. V., Seabra, V. S., Bastos, I. B., & Costa, E. C. P. (2020). A importância da análise espacial para tomada de decisão: um olhar sobre a pandemia de COVID-19. *Tamoios, 16*, 125–137. Disponível em: https://www.e-publicacoes.uerj.br/index.php/tamoios/article/view/50440. Acesso em: 22 out. 2020 (in Portuguese)

Cardoso, P. V., Seabra, V. S., Costa, E. C. P., Pimentel, J. R., & Machado, K. P. (2022). Análises espaciais dos casos de COVID-19 no Município de São Gonçalo (RJ). In: COVID-19: Análises e representações da pandemia no estado do Rio de Janeiro. Orgs: Vinicius da Silva Seabra, Phillipe Valente Cardoso. Editora Consequência. 1 ed. v.1. 2022 (in Portuguese)

Faria, R. M., Trentin, R., Petsch, C, Sccoti, A., Rizzatti, M, Lampert Batista, N., Cezar Spode, P. L, Bouvier Erthal, D., Vieira Da Silva, R., Quoos, J. H., & Turba Costa, I. (2021). Análise da infraestrutura médica da 4ª Coordenadoria Regional de Saúde do Rio Grande do Sul (RS) e dispersão da COVID-19. *Metodologias E Aprendizado, 3*, 114–137. https://doi.org/10.21166/metapre.v3i0.1314. Acessado 26 de dezembro de 2021 (in Portuguese)

Ibanez, P., de Sousa G. M., de Lucena A. J., de Farias H. S., de Oliveira L. D., da Rocha A. S. (2022). The COVID-19 Pandemic and Socio-spatial Inequality: A Study from the Metropolitan Area of Rio de Janeiro, Brazil. In: Saha, S., Mishra, M., Bhuimali, A. (Eds.), Economic and Societal Transformation in Pandemic-Trapped India. New Frontiers in Regional Science: Asian Perspectives, vol. 55. Springer, Singapore. https://doi.org/10.1007/978-981-16-5755-9_5

IBGE/SEA. (2018). Continuous Vector Cartographic Base of the State of Rio de January on Scale 1:25,000, 2018. Available at https://portaldemapas.ibge.gov.br/portal.php#mapa221075. Accessed 20 January 2020 (in Portuguese)

IBGE. (2021). Continuous Vector Cartographic Base of Brazil on Scale 1:250,000, 2021. Available at https://portaldemapas.ibge.gov.br/portal.php#mapa223417. Accessed 25 November 2021 (in Portuguese)

Instituto Brasileiro de Geografia e Estatística [IBGE] (2020) Estimativas de População, Brasil, 1.° de julho de 2020. https://www.ibge.gov.br/estatisticas/sociais/populacao/9103-estimativas-de-populacao.html. Accessed 6 July 2021(in Portuguese)

IBGE – Instituto Brasileiro de Geografia e Estatística. Estimativas da população residente para os municípios e para as unidades da federação brasileiros, com data de referência em 1° de julho de 2021. IBGE. 15 P. 2021. Disponível em: https://www.ibge.gov.br/estatisticas/sociais/populacao/9103-estimativas-de-populacao.html?=&t=o-que-e. Accessed 6 December 2021 (in Portuguese)

Lucena, A. J., Oliveira, L. D., Ibañez, P., et al. (2020). The geography of COVID-19 in Rio de Janeiro, Brazil: Conflicts, tensions, and challenges. In M. Mishra & R. B. Singh (Eds.), *COVID-19 pandemic trajectory in the developing world: Exploring the changing environmental and economic milieus in India* (pp. 51–69). Springer.

Méndez, R. (2023). Tiempos críticos para el capitalismo global: una perspectiva geoeconômica. Madrid: Espanha, 2023. Disponível em: https://revives.es/publicaciones/tiempos_criticos/. Acesso em: 15 ago. 2024.

Mclymore, A. (2021). Thousands of flights canceled globally as Omicron mars Christmas weekend. Reuters, New York. 25.12.2021. Disponível em: https://www.reuters.com/business/aerospace-defense/thousands-flights-scrapped-globally-omicron-mars-christmas-weekend-2021-12-24/. Acessado em: 26 de dezembro de 2021.

Monteiro, R. R., Angelotti, R.; Lautert, L. F. D. C.; Angelin, P. Eduardo et Portes, João. « "Rodovírus" ou "Caronavírus"? Mapas da Distribuição do COVID-19 na Região Sul do Brasil: Indícios da contaminação por rodovias », ***Confins*** [En ligne], 45 | 2020, mis en ligne le 08 juin 2020, consulté le 30 deciembre 2021. URL: http://journals.openedition.org/confins/28246; https://doi.org/10.4000/confins.28246 (in Portuguese)

Nitahara, A. (2021). Menos mortes de idosos por covid-19 indicam avanço da vacina. Agência Brasil, Rio de Janeiro. 18.06.2021. Disponível em: https://agenciabrasil.ebc.com.br/saude/noticia/2021-06/menos-mortes-de-idosos-por-covid-19-indicam-avanco-de-vacina. Acessado em: 26 de dezembro de 2021.

O GLOBO. Eduardo Paes quer carnaval sem distanciamento ou restrição de público. G1, Rio de Janeiro. 03.10.2021. Disponível em: https://g1.globo.com/rj/rio-de-janeiro/noticia/2021/10/03/eduardo-paes-quer-carnaval-sem-distanciamento-ou-restricao-de-publico.ghtml. Acessado em: 26 de dezembro de 2021 (in Portuguese)

OMS. (Organização Mundial de Saúde). World Health Organization. Data. WHO COVID-19 dashboard. Disponível em: https://data.who.int/dashboards. Acesso em: 11 ago. 2024.

OPAS. Brasil receberá as primeiras vacinas contra COVID-19 por meio do Mecanismo COVAX neste domingo. Brasília. 21.03.2021. Disponível em: https://www.paho.org/pt/noticias/21-3-2021-brasil-recebera-primeiras-vacinas-contra-covid-19-por-meio-do-mecanismo-covax. Acessado em: 26 de dezembro de 2021 (in Portuguese)

OPAS. (Organização Pan-americana da Saúde). "Histórico da pandemia de COVID-19". Disponível em: https://www.paho.org/pt/covid19/historico-da-pandemia-covid-19. Acesso em: 05 ago 2024.

OPAS. (Organização Pan-americana da Saúde). "Histórico da pandemia de COVID-19". Disponível em: https://www.paho.org/pt/noticias/5-5-2023-oms-declara-fim-da-emergencia-saude-publica-importancia-internacional-referente Acesso em: 11 ago. 2024.

Pereira, V. H. C., Rocha, G. C., Diniz, M. T. M., & Oliveira Jr., M. A. C. (2021). Análise de dependência espacial da taxa de mortalidade por COVID-19 nos municípios brasileiros. CONFINS, Revista Franco-Brasileira de Geografia. Dossiê Aspectos geográficos da pandemia de COVID-19. N.52. https://doi.org/10.4000/confins.40509. (in Portuguese)

Pinheiro, Lara e Garcia, Mariana. Pela primeira vez, mundo registra mais de 1 milhão de casos de Covid em 24 horas. G1, Rio de Janeiro. 28.12.2021. Disponível em: https://g1.globo.com/mundo/noticia/2021/12/28/pela-primeira-vez-mundo-registra-mais-de-1-milhao-de-casos-de-covid-em-um-unico-dia.ghtml. Acessado em: 29 de dezembro de 2021 (in Portuguese)

Ritchie, H., Mathieu, E., Rodés-Guirao, L., Appel, C., Giattino, C., Ortiz-Ospina, E., Hasell, J., Bobbie Macdonald, R., Beltekian, D., & Roser M. (2021). **Coronavirus Pandemic (COVID-19).** Published online at OurWorldInData.org. 2021. Retrieved from: 'https://ourworldindata.org/coronavirus' [Online Resource] Acessed 6 december 2021.

Schreiber, M. (2021). Rejeição de 70 milhões de doses da Pfizer por gestão Bolsonaro será novo foco da CPI da Covid. BBC News, Brasília. 09.05.2021. Disponível em: https://www.bbc.com/portuguese/brasil-57018138. Acessado em: 26 de dezembro de 2021 (in Portuguese)

SES-RJ. (2021). Daily epidemiological bulletin COVID-19, 2021. Available at https://coronavirus.rj.gov.br/boletins/. Accessed 20 December 2021.
Secretarias Estaduais de Saúde. Painel Coronavírus. Brasil, 2021. Available at https://covid.saude.gov.br/. Accessed 20 December 2021.
Sousa, G. M., Rocha, A. S., Farias, H. S., & Oliveira, L. D. (2022). A Baixada Fluminense e a COVID-19: A evolução na dinâmica da pandemia. In: COVID-19: Análises e representações da pandemia no estado do Rio de Janeiro. Orgs: Vinicius da Silva Seabra, Phillipe Valente Cardoso. Editora Consequência. 1 ed. v.1. 2022 (in Portuguese)
WHO. (2021). World Health Organization. Health Emergency Dashboard. WHO (COVID-19) Homepage. https://covid19.who.int/

Heitor Soares de Farias was awarded a Ph.D. in Geography from the Federal Fluminense University, and he is a Professor of the Postgraduate Program in Geography at the Federal Rural University of Rio de Janeiro. Works with Environmental-Urban Planning based on Geographic Climatology, Risk Analysis and Health Geography. He serves as a Member of the Integrated Laboratory of Applied Physical Geography-LiGA/UFRRJ.

Gustavo Mota de Sousa is Doctor in Geography from UFRJ (Rio de Janeiro, Brazil). He is an Associate Professor of Cartography and Remote Sensing at UFRRJ—Seropédica (Rio de Janeiro, Brazil) and Coordinator of the Integrated Laboratory of Applied Physical Geography (LIGA-UFRRJ)

Vinicius da Silva Seabra is Doctor in Geography from UFRJ (Rio de Janeiro, Brazil). He is an Associate Professor of Cartography and Geoprocessing at UERJ (State University of Rio de Janeiro) and a member of the DAGEOP group (Environmental Dynamics and Geoprocessing) at UERJ.

Phillipe Valente Cardoso is Doctor in Geography from UFRJ (Rio de Janeiro, Brazil). He is an Adjunct Professor of Cartography and Geoprocessing at UERJ (State University of Rio de Janeiro) and a member of the DAGEOP group (Environmental Dynamics and Geoprocessing) at UERJ.

Thiago Jeremias Baptista is Doctor in Geography from the Pontícia Universidade Católica do Rio de Janeiro (PUC-Rio), carried out a postdoctoral internship in Geography at the Graduate Program in Geography at the State University of Rio de Janeiro (UERJ). He currently works as a substitute professor of Human Geography at the Federal Rural University of Rio de Janeiro (UFRRJ), as well as Discipline Coordinator of the Cederj/UERJ Consortium. Currently, he conducts research articulating Economic Geography, Geopolitics and International Relations. In addition, he participates as a volunteer researcher at the Center for the Study of Fluminense Geography (NEGEF/UERJ/PUC-Rio) and as a researcher at the Research Group on Contemporary Economic-Spatial Restructuring at UFRRJ.

Chapter 11
The Agenda of South-South Cooperation and the Trajectory of Brazil's Foreign Policy 2003–2020: Rapprochement Between Africa and Latin America

André Santos da Rocha

Abstract South-South cooperation (SSC) is a polysemic and contradictory term, but it indicates closer relations between developing and underdeveloped countries (the so-called Global South). Since 2003, Brazil has adopted SSC as a strategy of its foreign policy and there has been a gradual growth in cooperation agreements and also in trade flows with countries in Africa and Latin America, but after 2011, there has been a significant decline in the number of projects and actions around SSC. This decrease is the result of changes in internal policies that adopt other orientation models, whether marked by disinvestment or abandonment by ideological positions called anti-globalist. This text aims to present a theoretical and analytical review of these changes in SSC over the last 20 years in Brazil, presenting an analysis of the situation and data that allow an understanding of the divestments and abandonment of SSC as a Brazilian foreign policy strategy.

Keywords Foreign policy · Geopolitics · South-South vooperation · Brazil

11.1 Introduction

Since the early 1990s, the world power order has undergone significant changes. This period was characterized by the end of the bipolar order and the emergence of a uni-multipolar order, which marked the transition from the Cold War period to the consolidation of a warmongering American power, and by the economic multi-centrality of different spheres and circuits of capitalist power (Hobsbawm, 2007), as well as the emergence of new power centralities that challenge the current world power order (Agnew, 2008).

A. S. da Rocha (✉)
Department of Geography, Federal Rural University of Rio de Janeiro (UFRRJ), Rio de Janeiro, Brazil
e-mail: asrgeo@ufrrj.br

M. Mishra et al. (eds.), *Contemporary Geopolitical Alignment in the Global South*,
https://doi.org/10.1007/978-3-031-81977-3_11

A myriad of authors such as Eric Hobsbawm (1995, 2007, 2008), Michel Hardt and Antoni Negri (2010, 2005), John Agnew (2008), David Harvey (2001), and Chomsky (2004) understand that this context is due to the predominance of North American hegemony, but also for its glimpses of crisis, where the emergence of other centers of power can be perceived based on the potential for association and cooperation perceptible in the globalized contemporary world system. We can highlight as recent facts the role of China competing with the United States, a technological race around 5G technology, and also the recent war in Ukraine, which highlights the fraying of the *Pax Americana* and the role of North Atlantic Treaty Organization (NATO) in the conflict with Russia.[1]

In this context of intense transformations in the global political scenario, there is a strengthening of complex economic and technical relationships between countries considered "peripheral and semi-peripheral," many of them organized around the so-called South-South cooperation (Mawdsley, 2017). The term used to describe this approach to South-South cooperation is intended to imply a more "gentle" or less "perverse" model than that established in the "North–South" economic and political interaction (Lima, 2005; Pautasso, 2010). However, Carlos Milani (2012, p. 2013) indicates that the distinction between these relationship/cooperation models "is fundamentally empirical." That is, the so-called "South-South" cooperation will not necessarily establish a relationship of "balanced relations," as this model often reveals profound ambivalences in practice (Mawdley, ibidem; Rocha, 2019). Thus, we cannot reduce the analysis to the discursive level of the practice of the countries surrounding the cooperation, but to its consequences for the participating countries. This implies, in our view, the social-environmental, territorial, and hegemonic developments of this cooperation—establishing a Geography of South-South cooperation that brings a new perspective to the world power order.

According to Maria Regina Soares Lima and Carlos Milani (2016), the term "South-South relations" or South-South Cooperation (SSC) is quite polysemic, as it refers to a variety of foreign policy actions. The rapprochement between countries of the Global South (Abdenur; Fonseca, 2013), in short, refers to mutual assistance in technical, economic, and political scenarios, however, full of contradictions. This type of diplomatic relations was born in 1955 with the "Bandung Conference," however, according to Diego Pautasso (2010, p. 49), the association between countries of the South emerged with greater force after 1970 when a conjuncture of reorganization of the world power was configured.

In this context, we can highlight the role of Brazil in the international scenario, which in the twenty-first century was based on the dynamics of cooperation as a pillar of its foreign policy, and which had a strong focus on the so-called South-South cooperation, in this context, the actions of cooperation with the countries of Latin America and Africa. Brazil has been involved in different forums, we can mention: the constitution of the BRICS and its financing bank; IBSA Dialogue Forum (India, Brazil, and South Africa); Preferential Trade Agreement (ACP) and

---

[1] As evidence of the paradigmatic crisis in the power relations scenario, it is inserted the speech of President Vladimir Putin on the breaking of the current world power order.

the MERCOSUR-SACU (Southern African Customs Union).[2] However, recently, in view of the changes in government, there have been reorientations, modifying the strategies of rapprochement with these countries.

This text presents notes on the characteristics of Brazilian foreign policy strategies between 2003 and 2022 and seeks to point out the comings and goings of actions centered on South-South cooperation from Brazil to Latin America and Africa. The analysis is organized around a literature review and the interpretation of economic data obtained from Comexstat (Brazilian foreign trade statistics database) and Brazilian technical cooperation from the Brazilian Cooperation Agency.

The text is divided into two parts. The first seeks to present the vision of Brazil and its consolidation as a regional power center. The second seeks to present the history of Brazilian foreign policy, with emphasis on cooperation initiatives for Latin America and Africa, presenting the differences and changes over the last 20 years.

11.2 Brazil, the New Dynamics of Global Power and Its Foreign Policy

Brazil is considered by many theorists as a strategic country in the new order of global power (Stuenkel, 2017). In addition to the characteristics of size of its territorial dimension and large resource capacity, which are considered strategic in the classical conception of geopolitics (Raffestin, 1993), we must also consider the relationships of proximity and influence that the country has played in recent years, especially for the countries of the Global South.

Daniel Flemes (2010) draws attention to Brazil's role in the organization and composition of the BRIC.[3] Through its participation, formulation, and composition of this group (Stuenkel, ibidem.), the country represents a point of tension in the balance of global forces. This is because they represent alternatives (financial, political, economic, and technical) for international development, which is largely fueled by the perspectives of South-South cooperation. Moreover, the Goldman Sachs Group pointed out in the early 2000s that Brazil, as an emerging country, was at that time becoming a good investment opportunity, comprising together with Russia, India,

---

[2] Trade agreement and existing tariff preferences between MERCOSUR (Southern Common Market, composed of Brazil, Argentina, Uruguay, Paraguay and Venezuela) formalized with SACU (Southern African Customs Union—Customs Union of Southern Africa, composed of South Africa, Namibia, Swaziland, Lesotho and Botswana). See: http://www.mdic.gov.br/index.php/comercio-exterior/negociacoes-internacionais/132-acordos-dos-quais-o-brasil-e-parte/1833-acordos-mercosul-sacu Access Date 07/19/2018.

[3] Acronym of the initials of Brazil, Russia, India and China. Created in 2001 by the financial group Goldman Sachs to designate countries that would represent good investment focuses due to high growth rates above the world average. Later, some specialists already indicated the addition of the letters S indicating the inclusion of South Africa (in English South Africa).

and China a group of countries that had attributes such as large territories, capabilities of natural resources and a large population that allowed to consolidate their economies in the forthcoming scenario (O'Neil, 2021).

Diego Pautasso (2010, p. 53) points out that "emerging countries are precisely the synthesis of international transformations, as they represent a change in the balance of power." If Brazil is an integral part of this context, the study of Brazilian South-South cooperation policy becomes relevant. However, the international cooperation agenda, as well as the strategies adopted, are directly linked to the position of an "internal political agenda" (Leite, 2011). In this way, the approach to peripheral and semi-peripheral countries as a geopolitical strategy is directly linked to its foreign policy perspective (Rocha, 2018).

Regarding the reading of Brazilian foreign policy, the rapprochement with Latin American and African countries is based on an agenda that has gained momentum in recent years, especially in the context after 2003, under the leadership of Luiz Inácio Lula da Silva (Visentini; Pereira, 2008). They gained prominence in the technical cooperation agenda through ABC projects—(Agência Brasileira de Cooperação—Brazilian Cooperation Agency) (Milani, 2017) and from an economic perspective through bilateral and multilateral agreements.

Despite the comings and goings of Brazilian foreign policy toward the countries of the Global South (Leite, 2011;), it is possible to think that already in the 1950s specific perspectives for Latin America and the African continent were constructed, especially in the *Pan American Operation (OPA in the Portuguese acronym)* of Juscelino Kubitschek (1958) in *the Independent Foreign Policy* (1961) of the Jânio Quadros government. Paulo Visentini and Pereira (ibidem) emphasize that there was no continuity of this policy during the Cold War, especially in the context of the military dictatorship. The scenario in which there is a systematic decline in Brazilian interest in relations with Africa and Latin America.[4] This lack of interest extends to the early 1990s with the Collor government.

Lima (2005, p. 11) complements this issue by addressing the prestige and autonomist character that Brazilian foreign policy assumed under the governments of Fernando Henrique (1995–2002) and Luiz Inácio Lula da Silva (2003–2010). In both administrations, cooperation with African and Latin American countries was resumed and intensified in the face of the current situation, as an alternative to the crisis affecting the United States and the countries of the European Union (COSTA Lima; MELO 2010).

If we think about the most recent situation, within the timeframe of the last 20 years, we have received different variations of emphasis on the construction of this Brazilian foreign policy agenda. Between 2003 and 2010—we had the government of Luiz Inácio Lula da Silva, between 2011 and 2016—the government of Dilma Rousseff, between 2016 and 2018—the government of Temer, and between 2018 and

[4] It is important to highlight that in the period that marks the military dictatorship in Brazil, the recognition by the Brazilian government of the independence of countries on the African continent and, respectively, the opening of embassies in these countries and the bilateral approximation with Latin American countries must be considered.

Table 11.1 Comparison of Brazilian governments between 2003 and 2022 and their foreign policy perspective

Ruler	Management period	Political party	Guidance	Foreign policy action/ perspectives
Luiz Inácio Lula da Silva	2003–2010	PT (Workers' party)	Left	Pragmatic globalist multilateralism / focus on high-intensity South-South cooperation /foreign policy based on "autonomy through diversification"
Dilma rousseff	2011–2016	PT (Workers' party)	Left	Globalist multilateralism/ focus on south-south cooperation – low intensity/ "continuity containment foreign policy"
Michel temer	2016–2018	MDB	Center-right	Multilateral pragmatism, with the return of relations with the European Union and the United States
Jair Bolsonaro	2019–2022	PSL / PL	Far-right	Anti-globalism/strong bilateralism tendency with strong adherence to North American interests

Source Elaborated by the author

2022—the government of Jair Messias Bolsonaro. However, it is not easy to associate all the links between the internal and external policies of countries (Fearon, 1998). In the Brazilian case, we can see that each government, with a different internal political orientation, directed foreign policy actions differently (see Table 11.1).

The two terms of Luiz Inácio Lula da Silva (Lula) (2003–2010) were marked by a profound change in the scenario of political leadership in the country. For the first time since the re-democratization (1985),[5] a government declared to be of the left-wing assumed national guidelines. In this context, there is a great impulse in the Brazilian cooperation agency as well as in the promotion of the so-called South-South cooperation. Agreements with countries of different shades are being intensified, without abandoning the pragmatism associated with consolidated partners such as the United States and the European Union. However, there is a rapid growth of agreements and commercial exchanges with China, Russia, and India (which would become a partner in the BRIC), as well as technical and budgetary advances in cooperation with countries in Latin America and Africa. Vigevani and Cepaluni (2007) call these foreign policy actions "autonomy through diversification."

---

[5] This period refers to the resumption of civil governments after a long period of military dictatorship from 1964 to 1985. Direct elections only took place in 1989 for the election of Fernando Collor de Mello, who would preside over Brazil between 1990 and 1992.

It suffices to remember that in 2010 [the last year of Lula's administration], the proportional amounts invested in international technical cooperation by region of the world were US$19,844,882.00 (Africa); US$14,009,188.00 (Latin America and the Caribbean); and US$2,158,840.00 (Europe, Asia, Oceania, and the Middle East)[6] (Rocha, 2018). This budget made clear the priorities in the area of South-South cooperation. Also noteworthy is Brazil's participation in numerous international forums, as well as its prominent leadership role in some agendas.

Dilma Rousseff, who took over the presidency under the social and economic legacy of Lula's government, has a full term (2011–2014) and an incomplete one (2015–2016). According to Cornetet (2014), she has in practice a "foreign policy of containment in continuity." During this period, Brazil advanced in international forums constituted as BRICS and continues with international cooperation actions in the South-South agenda, but since 2014 cooperation projects have already shown a decline in the number and volume of investments, as well as a reduction in propositional actions such as international travel and diplomatic missions. Hence the term "containment on permanence". In support of this issue, Cornetet (2014: 140) points out that:

> it was found that during Dilma's government, compared to Lula's, there were fewer presidential trips - about 30% less if we compare it to Lula's first term, about 50% less if we compare it to the second term - foreign policy initiatives were less numerous and more timid, the training of new diplomats decreased - to about half of that of Lula's first term and about a quarter of that of the second term - and the establishment of new diplomatic posts slowed down

The government of Michel Temer (2016–2018) was born with a strong institutional rupture as a result of a legal-parliamentary coup that interrupted Dilma Rousseff's second term. In terms of foreign policy as a center-right policy, it marked the appointment of Chancellor Aloísio Nunes, who, among other measures, adopted the suspension of Brazil's participation in the BRICS, as well as a shift in the centralization of actions toward a greater rapprochement with the European Union and the United States. There is a multilateral pragmatism because it reinforces the perspective of direct return of Brazilian actions abroad, differentiating itself from previous governments which brought the centrality of South-South cooperation, which, in theory, would not necessarily require an immediate potential gain. On the other hand, it seeks to resume trade relations with Mercosur.

The government of Jair Messias Bolsonaro (2019–2022) is characterized by a shift to the right wing, both in domestic and foreign policy. There was a clear tendency to abandon the legacy built in previous governments around South-South cooperation in favor of a deliberately controversial agenda. The presence of Chancellor Ernesto Araújo stands out, who presented and supported anti-globalist actions and a strong adherence to US interests (especially under the leadership of Donald Trump), following the guidelines of extremist and dictatorial countries. In the Latin American

[6] Check Corresponding Charts—South America and the Caribbean http://www.abc.gov.br/Content/ABC/imagens/americas_financeiro.png ; and Asia, Europe, Oceania and the Middle East http://www.abc.gov.br/Content/ABC/imagens/asia_financeiro.png.

context, it was marked by a deep split, with a strong denial of belonging to South American countries linked to the left and extreme left, which materialized in the abandonment of UNASUR for the constitution of PROSUL.[7] We emphasize that the actions of South-South cooperation did not stop, but they decreased in volume and in financial contributions. The government period was still marked by problems with China and various European countries, due to impersonal speeches (postings on the Internet) and official pronouncements by the former president, his children, and his chancellor. There were spotlights on the crisis installed during the most problematic period of COVID-19, in which some leaders made xenophobic speeches against China, and gained prominence (Ibañez, 2020).

11.3 Some Clues of the Detours of Brazilian Cooperation on the South-South Prism

The definition of South-South Cooperation (SSC) is quite controversial, especially because of the wide range of actions that this type of cooperation can involve (Mawdsley, 2012). In any case, the actions of the United Nations (UN) take as the general scope of SSC the indications of the Plan of Action of Buenos Aires (1978).[8] These guidelines include policies that will guide technical and economic cooperation between countries to promote, among other things: [a] the self-reliance of developing countries and their creative capacity; [b] the formulation of development strategies focusing on the sharing of experience, technical resources, and other inputs; [c] the greater capacity of countries to participate in international economic activities than some countries; and [d] the technological capacity of developing countries as well as the promotion of the adaptation of technologies and skills to the specific characteristics of each country.

In the Brazilian case, international cooperation based on a South-South approach, which is a signatory to the general guiding principles of this agenda: horizontality, non-conditionality, non-commercial binding, non-interference, and not indifference (Ferreira and Moreira, 2018) will gain prominence and emphasis between 2003 and 2014. South-South cooperation was understood as a geopolitical strategy (Rocha, 2019), as it allowed Brazil, as a country of great proportions (compared to the most impoverished countries) of technology and innovation in certain sectors, to gain prominence with countries in Latin America and Africa. Technical cooperation actions gained prominence, which is a modality that involves the exchange of innovative technologies, procedures, and methods in different sectors. These actions also

[7] UNASUR—The Union of South American Nations—is an inter governmental regional organization set up by Hugo Chavez to counter act the influence of the United States in the region where as the PROSUL—The Forum for the Progress and Integration of South America—was created by the former president of Chile, Sebastián Piñera and the former president of Colombia, Iván Duque to counter balance the UNASUR seen as pro-left.

[8] For more information consult: https://www.ilo.org/brasilia/temas/south-south/WCMS_660549/lang--pt/index.htm. Access 07 Feb 2023.

include the exchange of experiences in management and administration, as well as in the training of people.

The main feature of technical cooperation in the South-South prism in Brazil has been *demand driven*. This is when other countries ask Brazil to carry out cooperation activities. By identifying the existing capacities and potentials in the country. We have already highlighted in other texts that this clearly reveals Brazil's hegemonic potential, especially in the areas of agriculture, education, and health (Rocha, 2018). Moreover, given the complex forms in which South-South cooperation takes place, it is impossible not to consider the diplomatic role in economic and financial rapprochement. Therefore, it is also possible to think that import and export flows between countries reveal, to a certain extent, a relationship of alignment with foreign policy guidelines. In the end, this reveals ambiguities and creates asymmetries in the field of South-South cooperation, which is not so different from North–South cooperation (Mawdsley, 2017).

As we have seen, the direction of Brazilian domestic politics has changed significantly in the last 20 years, from progressive left-leaning governments between 2003 and 2016, to right-wing governments and a strong adherence to liberal policies (2016–2022). The changes have significantly altered the rapprochement with the countries of the Global South, especially when we think about the rapprochement with the countries of Latin America and Africa.

As an example, in the field of cooperation, one can note the gradual decline in the number of agreements and cooperation projects developed for Africa. In particular, a continent that became strategic in the geopolitical maintenance of Brazil as a country with a voice of global reach, through the so-called prestige diplomacy (Visentini, 2010), in which the relationship could develop in support of African countries in the diplomatic rounds of international forums. Among the African countries, the PALOP (Portuguese-speaking African countries) stand out, such as Angola, Mozambique, Guinea-Bissau, Cape Verde, and São Tomé and Principe. These countries have traditionally been close to Brazil and are the countries with the highest number of projects in terms of volume. However, they have been gradually declining since the last Lula administration in 2010 (See Fig. 11.1).

According to the data of the Brazilian Cooperation Agency, the number of projects developed in the PALOP countries decreased even under the government of Dilma Rousseff, which can be understood as a foreign policy of "continuity in containment," indicating that even with the same internal orientation, the intensity of technical cooperation actions is cooled. There is a slight increase in the short period of the Temer government (2016–2018) and a direct decrease in cooperation actions in the government of Jair Messias Bolsonaro (2019–2022).

We highlight here the social and hegemonic importance of technical cooperation, which refers to the ability to share experiences, and highlights the level of Brazilian development in certain sectors. Its decline reveals not only the disinvestment in the foreign policy sector but also in the Brazilian prestige and influence capacity, since it opens space for other countries that are also interested in these African partners, as in the case of China (Abdnur, 2014; Ajakaiye; Kaplinsky, 2009). It is possible to identify the same downward trend in relations, with an emphasis on South-South

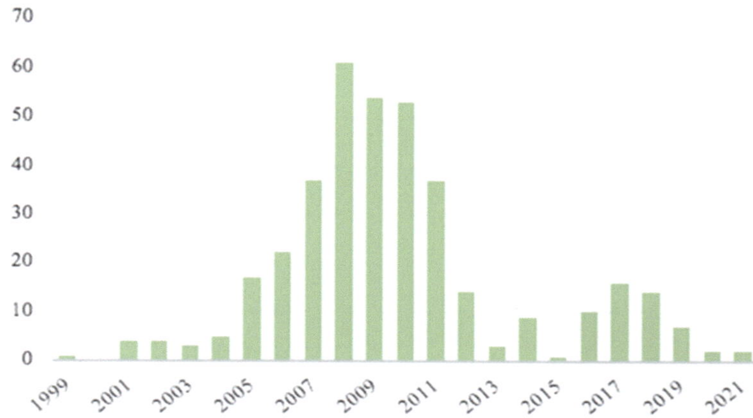

Fig. 11.1 Number of projects carried out in the PALOP (1999–2021) (*Source* Brazilian Cooperation Agency (Ministry of Foreign Affairs), Prepared by Mariana Herreira)

relations, in the decline of Brazilian exports and imports to Latin America after 2012. (See Fig. 11.2).

Figure 11.2 expresses the same upward trend, highlighting a significant increase in the volume of exports and imports between 2003 and 2010, reaching the highest export peak in 2013, with a decline until 2016, when it shows a small recovery. Although not necessarily revealed as an extension of the SSC, trade is indicative of the approximation of these countries. And it is very clear in the graph that it is in the Lula governments (2003–2010) that there is the greatest exponential growth in the volume of imports and exports, decreasing from the Dilma governments (2011–2016), resuming, even timidly, in the Temer governments (2016–2018).

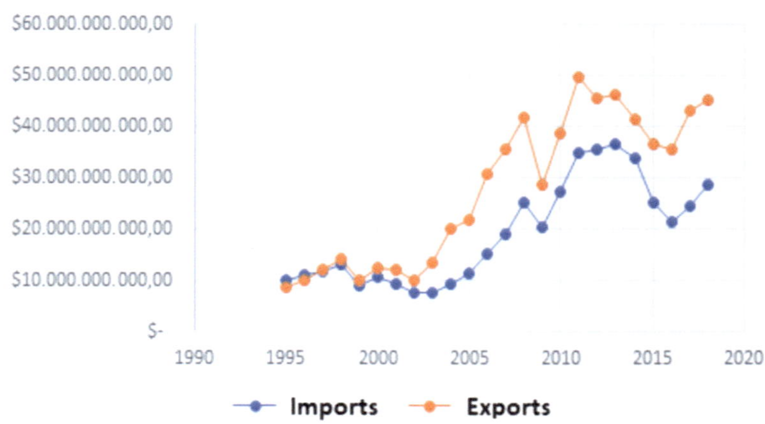

Fig. 11.2 Brazilian trade flow with Latin American countries between 1995 and 2019 (*Source* Comextat, 2020 (Historical series), Prepared by Maria Fernanda Arcanjo)

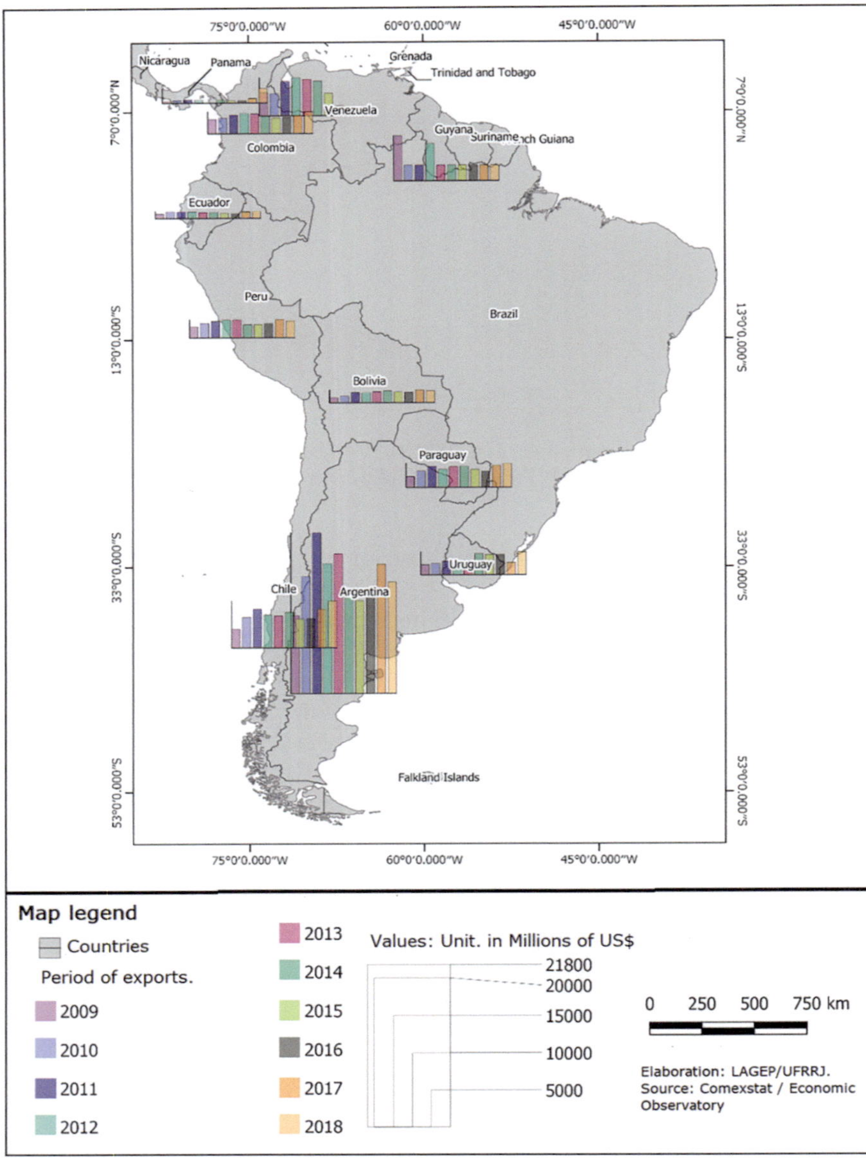

Fig. 11.3 Map with the histogram of Brazilian exports to South America (2009 2018) [Created by the author. *Source* ComexStat (2018)]

11.4 Final Considerations

At first glance, South-South cooperation appears to be a direct alignment of domestic strategies, but its model, the nature of its implementation, and its diplomatic and economic returns must be considered in the long term. Either to promote the image of Brazil as a regional-world potential or to consolidate economic partnerships (although it is not the SSC Institute).

The political history of Brazil in the 20 years showed that the adoption of SSC strategies had a positive impact on the Brazilian image in the world, and also on economic growth with countries in Latin America and Africa, but the internal changes and political instabilities that characterize the country since 2015 have brought a significant decrease with partners from the Global South. This decrease comes from the disinvestment in technical cooperation actions to the detachment from the model, understood as an "ideological" stance (especially in the Bolsonaro government). Strategies to maintain the SSC as a pillar of Brazilian foreign policy should be treated as a geopolitical element and not as an ideological dimension, given the capacity of this model to establish the Brazilian commitment to development, solidarity, and maintenance of technical hegemony in important sectors such as education, agriculture, and health.

Likewise, the resumption of dialogue at the regional level and Brazil's leadership in forums in the Global South are extremely important, given their importance for the economic vitality of exports. In a broad sense, we can see that the abandonment of South-South cooperation took place in three stages. The first in the government of Dilma Rousseff, with a "foreign policy of containment in continuity" has a decrease in the number of projects and cooperation agreements, as well as the intensity of exchanges with countries of the Global South. The second with Michel Temer (2016–2018), who in a short period bet on multilateral diplomacy, under the neoliberal guidelines, which continues disinvestment in South-South cooperation projects, but tries to strengthen regional trade with countries of the Southern market. Finally, the third moment of complete abandonment of the SSC in the Bolsonaro government (2019–2022), for ideological reasons, creates closer ties with the United States and promotes, through its chancellery, the abandonment of regional forums of the Global South.

Indeed, studies on the impacts of disinvestments in the field of South-South cooperation need to be improved, the construction of agendas on Brazilian foreign policy now opens a new chapter. Once again, Luiz Inácio Lula da Silva becomes president for the term between 2023 and 2026, but it is still too early to indicate a resumption of SSC as a hegemonic and geopolitical Brazilian projection strategy, as well as the expansion of partnerships with the countries of the Global South.

Acknowledgements We would like to thank FAPERJ for funding the Graduate Program in Geography through the public notice to support graduate programs in Rio de Janeiro. We also thank CNPq and FAPERJ for the scholarships granted to Mariana Herreira and Maria Fernanda Arcanjo, who participate in research projects with our laboratory, working in the preparation of maps, graphs, and various analyses.

References

Abdenur, A. E. (2014). China and the BRICS development bank: Legitimacy and multilateralism in South-South cooperation. *Institute of Development Studies Bulletin, 45*, 85–101.

Abdenur, A. E., & Fonseca, J. (2013). The North's growing role in South-South cooperation: Keeping the foothold. *Third World Quarterly, 34*(8), 1475–1491.

Agnew, J. (2008). A nova Configuração do Poder Global. *RevistaCaderno CRH, Salvador, 21*(53), 207–219, Maio/Ago.

Ajakaiye, O., & Kaplinsky, R. (2009). *China in Africa: A Relationship in Transition. European Journal of Development Research (EJDR) 21*, 479–484.https://doi.org/10.1057/ejdr.2009.30

Chonsky, N. (2004). *O Império Americano - HegemoniaouSobrevivência.* São Paulo: Campus

Cornetet, J. M. C. (2014). Política externa de Dilma Rousseff: contençãonacontinuidade. *Revista Conjuntura Austral, 5*(24), 111–150. https://seer.ufrgs.br/ConjunturaAustral/article/view/47628/30012. Accessed 10 Jun 2018

Costa Lima, M. F., Melo, V. D. S. (2010). Revalorização do lugar da África: Política de desenvolvimento e as relações Sul-Sul no Governo Lula da Silva. SÉCULO XXI, Porto Alegre, v. 1, n. 1, p. 127–155, jan-dez

Ferreira, P., & Moreira, A. (2018). Cooperação Tecnica Brasileira para África (2003–2013): Expansão orçamentária e limitações Institucionais. *BPC Policy Brief. 6*(01), Janeiro–Abril

Flames, D. (2010). Brasilnainiciativa BRIC: Soft balancing numaordem global emmudança? *Revista Brasileira de Política Internacional, 53*(1), 141–151.

Hardt, M., & Negri, A (2010). Império. 9ªed. Rio de janeiro: Record.

Hardt, M., & AMultidão, N. (2005). Guerra e democraciana era do império. Rio de Janeiro: Record.

Harvey, D. (2001). Spaces of capital: Towards a critical geography, nova york, routledge.

Hobsbawm E Era dos Extremos. (1995). São Paulo: Cia das Letras.

Hobsbawm, E. (2007). Globalização, democracia e terrorismo. São Paulo: Companhia da Letras.

Hobsbawm, E. (2008). *On Empire. America, war, and global supremacy.* New York: Pantheon Books.

Ibañez, P. (2020). Geopolítica e diplomaciaem tempos de Covid-19: Brasil e China no limiar de um contencioso. *Espaço e Economia.* https://doi.org/10.4000/espacoeconomia.13257

Leite, P. S. (2011). OBrasil e a cooperaçãosul-Sul emtrêsmomentos de política externa: osgovernosJânio Quadros/João Goulart,Ernesto Geisel e Luiz Inácio Lula da Silva. Brasília: Fundação Alexandre de Gusmão.

Lima, M. R. S. (2005). A política externa brasileira e osdesafios da cooperação Sul-Sul. *Revista Brasileira De Política Internacional, No, 48*(1), 24–59.

Mawdsley, E. (2012). *From recipients to donors: The emerging powers and the changing development landscape.* Zed Books.

Mawdsley, E. (2017). Development geography 1: Cooperation, competition and convergence between 'North' and 'South.' *Progress in Human Geography, 41*(1), 108–117. https://doi.org/10.1177/0309132515601776

Milani, C. (2012). Aprendendo com a história: críticas à experiência da Cooperação Norte-Sul e atuais desafios à Cooperação Sul-Sul. *Card.CRH, Salvador, 25*(65), 211–231. Maio/Ago. https://doi.org/10.1590/S0103-49792012000200003

Milani, C. (2017). *ABC 30 anos: história e desafios futuros.* Brasília: Agência Brasileira de Cooperação.

O'Neil, J. (2021). Building better global economic BRICs. Global Economics Paper No: 66. London, Goldman Sachs. Available at: https://www.goldmansachs.com/insights/archive/archive-pdfs/build-better-brics.pdf. Acessed 01 February 2023

Pautasso, D. (2010). Fortalecimento das relações sul-sul: Estratégia e realidade para os países emergentes. *Revista Conjuntural Austral, 2*(8), out-nov, pp. 48–62.

Raffestin, C. (1993). *Por Uma Geografia do Poder*. São Paulo: Ática

Rocha, A. S. (2018). A política externa e geopolíticabrasileirana era Lula: umaanálise dos acordos de cooperação técnica para a África. In: A. A. Rückert, A. C. Silva, & G. V. Silva (Eds.), Geografia

Política, Geopolítica e Gestão do Território: integraçãosul-americana e regiõesperiféricas. Porto Alegre: Editora Letra1, pp. 131–144.
Rocha, A. S. (2019). Cartografia e geopolítica das relações sul-sul: Sobre a cooperação técnica Brasil-África. *Confins* [on lline], 39. https://doi.org/10.4000/confins.17327
Stuenkel, O. (2017). *BRICS*. Paz e terra.
Vigevani, T., & Cepaluni, G. (2007). A política externa de Lula da Silva: aestratégia da autonomia pela diversificação. *Contexto Internacional, 29*(2). https://doi.org/10.1590/S0102-85292007000200002
Visentini, P. F., & Pereira, A. D. (2008). A política africana do governo Lula. Disponível em https://www.ufrgs.br/nerint/folder/artigos/artigo40.pdf. Acessed 14 Jan 2019
Visentini, P. F. (2010). Cooperação Sul-Sul, diplomacia de prestígio ou imperialismo "soft"? as relações Brasil-África no governo Lula. *Revista SÉCULO XXI*, Porto Alegre, V. 1, Nº1, pp. 65–84, jan-dez.

André Santos da Rocha holds a PhD in Geography from the Federal University of Rio de Janeiro (UFRJ) and a Postdoctoral Internship in Public Policies and Human Training from the Universidade do Estado do Rio de Janeiro (UERJ). He is an Associate Professor of the Department of Geography at UFRRJ and Permanent Professor in the courses of Master's and Doctorate (PhD) in Geography at UFRRJ. He is the coordinator of LAGEP/UFRRJ and works in the areas of Political Geography, Economics, and Urban-Regional approaches. He is the editor of Continents—a journal linked to the post-graduation program in Geography at UFRRJ. Presently, he is the investigator of produtivided—Jovem Cientista do Nosso Estado (Young Scientist our State) at FAPERJ (Research Support Foundation in the State of Rio de Janeiro)

Chapter 12
Traditional Territory in the Upper Amazon: The Consolidation of an Ethnographic Tourism

Wladimir Mejía Ayala and Júlio César Suzuki

Abstract Since 1970s, the vertiginous advance of the oil industry in the High-Amazonia and its socio-environmental impact led the indigenous peoples to organize themselves politically to resist changes in the daily life of the communities. During this process, tourism is developed in their territories, as an activity that goes beyond the economy, toward the conservation of the environment and culture, toward the protection of life itself. From the case of the Sápara people of Ecuador, based on their cultural reconstruction and their adaptation to new living conditions, we will evaluate the meaning of ethnographic tourism in the consolidation of the traditional Sápara territory in the Upper Amazon.

Keywords Indigenous territoriality · Spatial practices · Human-environment relations · Amazonian indigenous communities

12.1 Introduction

Comprising nearly 100,000 members, divided into 28 ethnolinguistic groups, the Sápara were one of the most important Indigenous groups in the Amazon (Trujillo, 2001), until their number was reduced during colonization, evangelization, the exploitation of rubber, and due to diseases, slavery, and forced displacement to which they were subjected. The ancestral territory of the Sápara people extended between the Napo and Tigre rivers, from the Andean foothills to the Marañón river basins. In 1941, the border war that ended with the signing of the Rio de Janeiro Protocol divided the Sápara territory between Peru and Ecuador (Fig. 12.1).

In Ecuador, the Sápara were culturally mistaken for the Kichwa and Achuar ethnic groups, being considered extinct in 1970 (Costales & Costales, 1975). According to

W. Mejía Ayala (✉)
Pedagogical and Technological University of Colombia, Tunja, Colombia
e-mail: wladimir.mejia@uptc.edu.co

J. C. Suzuki
University of São Paulo, São Paulo, Brasil

M. Mishra et al. (eds.), *Contemporary Geopolitical Alignment in the Global South*,
https://doi.org/10.1007/978-3-031-81977-3_12

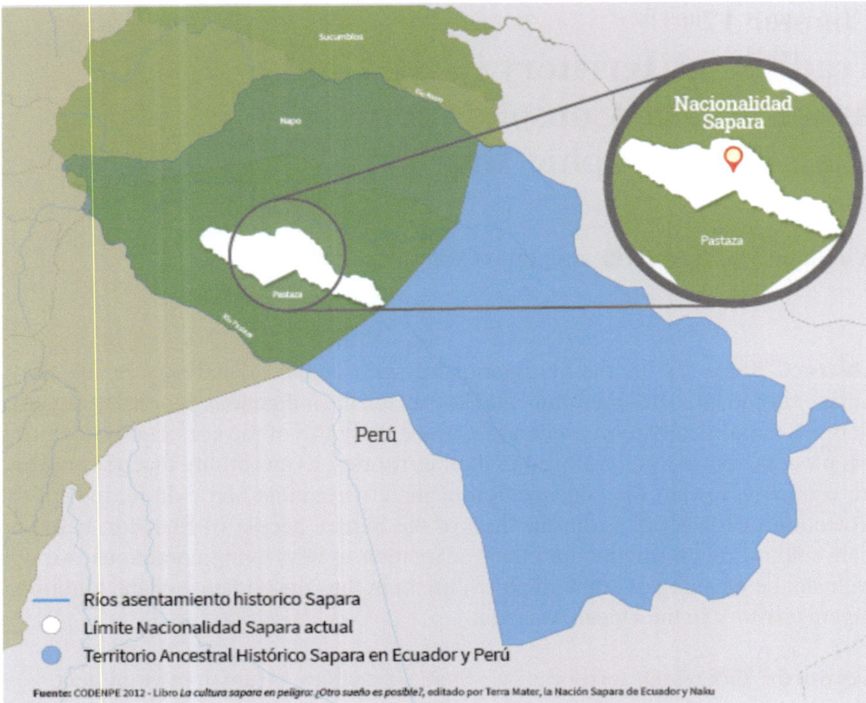

Fig. 12.1 Sápara ancestral territory past and present (*Source* Castillo, Mazabanda, Félix, Melo, Moreno de los Ríos, Narváez, Páez & Ushigua, 2016)

Bilhaut (2007), this responded to a camouflage strategy to be forgotten by their traditional enemies while benefiting from the advances of their political organizations. It was not until the 1990s, at the same time as the arrival and development of the oil industry in that part of the country, that the Sáparas of Ecuador reappeared, claiming their ethnicity and jurisdiction over a 370,000-ha territory in the current province of Pastaza, between the parishes of Río Tigre, Montalvo, and Sarayaku, in a mega-biodiverse area of difficult access. The closest route is 120 km away, in a seven-day walk (Pérez, 2005); the Conambo and Pindoyaku rivers representing the limits of this territory are not navigable, and their confluence, which gives birth to the Tigre River on the border with Peru, continues to be controlled by the armies of both countries. This space remains hypothetically closed and, in a certain way, at the margin of the market economy. It is accessed only by air after a 45-min flight, at a cost of around 500 US dollars.

Despite the complexity of the terrain, incursions of tourists began in the mid-1990s and especially from the year 2001, when the Sápara were proclaimed by Unesco as a Masterpiece of the Oral and Intangible Heritage of Humanity. Since then, and despite the difficulties experienced by travel organizers such as distance, mobility, price, tourist groups from 2 to 8 people visit this space with relative frequency, for at least ten days.

In this context, the goal was to understand the origin and foundations for the spatial and cultural opening of the Sápara, allowing the arrival of tourists to their territory. The tourism represents to them more than what proposes in its definition: a new economic activity to generate income and diversify other pre-existing activities (Duhamel, 2003), a strategy to preserve their territory in the face of the threat posed by oil extraction, reaffirming their presence and authority and, consequently, guaranteeing the recreation, the conservation, transmission, and even the reinvention of their culture.

Thus, the objective of this analysis is to demonstrate, regarding the case of the Sápara people of Ecuador, that the origin of tourism in the Upper Amazon is directly and indirectly related to extractivism, particularly with the development of the oil industry.

This work consists of five parts: the first, Methodology and materials, describes the two phases in which the project was developed. It is a qualitative methodology of a descriptive and explanatory nature based on ethnography, geohistory, and cartography. The second, The Sápara people: A culture in recreation, narrates the process followed by the Sápara people to reconstruct themselves culturally and be identified based on their original cultural traits and the cultural traits introduced by members of other ethnic groups. In the third, a look at the Sápara ancestral territory, the problems and challenges that the Sápara people have faced during the consolidation of their territorial process are pointed out. The fourth, the limits of the territory, presents how the Sápara have adapted to the new living conditions in the face of the threat that means the entry of the oil industry, which is related to a sedentary lifestyle in a limited space and to the resources the territory can offer. The fifth part, titled Final considerations, origin and foundations of tourism in the Amazon? Represents the synthesis of the whole work and explains the origin and the foundations of tourism in the Upper Amazon.

12.2 Methodology and Materials

This qualitative study is one of a descriptive and explanatory nature developed in two phases. In the first phase, there was a review of secondary information. Here, it is possible to highlight, besides institutional and historical documents, the previous works carried out based on the case of the Sápara people, namely, the thesis by Maximiliano Viatori (2005), "The language of "authenticity": Shifting constructions of Zápara identity, the politics of Indigenous representation, and the state in Amazonian Ecuador," developed in the Department of Anthropology of the University of California in which the author explores the practices of self-representation of some indigenous leaders from the Sápara Nation of Ecuador (NASE). Another work to consider is Anne-GaëlleBilhaut's Ph.D. thesis in Ethnology (2007), "The awakening of the immaterial. The dreamlike production of the heritage of the Zápara Indians

(Upper Amazon)," conceived at the University Paris Nanterre, which shows how the Sápara people gain knowledge of their past through dreams, playing a fundamental role in their daily lives. The history of these people is constructed with the help of dream experiences. In the same way, these dreams are used strategically in the present. Finally, the master's thesis of Agnes Müller (2011) developed at the University of Vienna, based on the contact between the Sápara and Shuar languages as well as the work of the ethnolinguist Carlos Andrade Pallares, written during the process of preparation and presentation of the UNESCO proclamation application dossier, between 1999 and 2001, and published in 2001, are to be considered.

The second phase consisted of ethnographic fieldwork implemented between July and October 2010, based mainly on participant observation. Initially, exchanges were made with members of the Sápara Nation of Ecuador (NASE), in the city of Puyo, in the Pastaza province, where the headquarters of that political organization are. In that city, exchanges were made as well, with other actors involved in different ways in the Sápara territorial process. Then, two incursions were made into the Sápara ancestral territory: the first, accompanied by a political leader, his son, and two employees of a local non-governmental organization; the second, with a group of tourists. It was possible to collect audio and photographic information, georeference places and events, observe actively at different moments, and keep a field diary. The result is presented as a geohistorical, according toBraudel (Ribeiro, 2017), and ethnographic study with an essential cartographic contribution.

12.3 The Sápara People: A Culture in Recreation

In the Sápara area live about a thousand inhabitants from three different ethnic groups: Sápara, Achuar, and Kichwa, distributed in 23 communities united in the political organization Sápara Nation of Ecuador (NASE). Although there are still tensions between the inhabitants, most of them recognize themselves as part of this nationality:

> Although the Sápara have grown through the union with the Kichwa, Achuar, and Shiwiar, the communities identify themselves as Sápara, and regardless of how much Sápara blood is in each person, all have embraced their culture and way of life, adding their characteristics to it (Bartolo Ushigua, former president of the NASE. Quoted by Rodríguez, 2007).

In Bartolo's speech, it can be confirmed what Claval (2012) pointed out, that culture is largely composed of words that translate what is perceived as real, structuring and organizing it. According to the author, these signs describe places, life, and all beings or techniques, but they also prescribe, that is, they do not indicate what is, but what should be. In this sense, the sense of identity or connectedness among the inhabitants of the Sápara territory, if it exists, is not the result of a common historical relationship and is not rooted in the practice and transmission of a common culture, but in the attachment to the geographical space. However, when there is coexistence, there is inevitably an exchange between the inhabitants: certain practices and knowledge are shared.

In fact, when the inhabitants have to revisit and later transmit practices and knowledge handed down to them by their ethnic group of origin, they adapt them to the needs of their time and to the characteristics of their environment and social circumstances. Thus, all the inhabitants of the communities play an important role in the revival, transmission, and production of a culture that becomes common to them. What we know today as the Sápara people is that they have managed not only to recover the socio-cultural references of their ancestors but also to introduce, adapt, and create new references. Consequently, the inhabitants of the Sápara ancestral territory are not only the summary of the Sápara, Kichwa, and Achuar cultural elements but also the representatives of this summary and its synergy.

12.4 A Look into the Sápara Ancestral Territory

An ancestral territory is a place that has been continuously inhabited by the ancestors of the people who inhabit it today and exercise authority and jurisdiction over it. This place does not answer to modern political-administrative boundaries but is demarcated and defined by the presence of a traditional society and by the way that society inhabits it, i.e., in terms of tradition. Although tradition is inherent in any society, the term is used to distinguish what is called a "traditional society"" from a Western or modern society.

In the past, the inhabitants of the Sápara ancestral territory were still semi-nomadic. Their movement throughout their land could last for significant periods, so inhabited places were difficult to locate. Their territory began to consolidate in the eyes of the State when seeing themselves affected by oil exploration and exploitation projects they realized that the ability to influence oil policy depended, first, on their degree of political organization (Fontaine, 2003) and second, on the fact of being present in places of interest to oil companies.

Concerning the first point, until the creation of the Sápara Nation of Ecuador (NASE) in 2009, the inhabitants of the Sápara territory belonged to different political organizations: Organization of the Indigenous Peoples of Pastaza (OPIP), Evangelical Indigenous Association of Pastaza for the Amazon Region, Union of Centers of the Zápara Territory of Ecuador, Organization of the Zápara Nation of Ecuador, National Zapara Association of the Pastaza Province of Ecuador and Zápara Nation of Pastaza; and they maintained internal territorial disputes and more or less violent power struggles. However, they consolidated strategic alliances with national political organizations such as the Confederation of Indigenous Nationalities of Ecuador and the Confederation of Indigenous Nationalities of the Ecuadorian Amazon; they consolidated regional alliances as the Coordinator of the Indigenous Organizations of the Amazon Basin, for the recognition of their rights, bilingual education and the legalization of their territories.

Regarding the second point, the inhabitants dispersed all over the territory, consolidating a network of settlements with houses and other edifications built along the Conambo and Pindoyaku rivers, making their limits less fluid. The first community founded was Conambo in the early 1970s, when some Achuar and Kichwa families, motivated by job opportunities and access to the money provided, gathered around the first airstrip that the Western Oil Company built, generating significant changes at the socio-cultural level:

> Oil companies produced substantial changes in the lives of the people of the Sápara territory. It was terrible because everyone got hooked on the company, and the money management for the world of nationalities is very different from ours, everything they earned went to buy recorders, they were fashionable; that is, the concept of saving does not exist. [...] I worked in that company (Western) when I was a guambra (teenager). At the time, it was the one who bought the biggest recorders that began to change its identity, like globalization today. There was a mini-globalization in the sense that the tape recorder was introduced as well as fashion, owning the tape recorder was great, just like previously having the shotgun, it was a symbol of power at the time of evangelization (Excerpt from interview with Carlos Duche. Puyo, October 5, 2010).

At the beginning of 1980, during the attempts by other oil companies to enter the Sápara territory, new communities were founded, including Alto Corrientes, Jandiayaku, and Masaramu, then the communities of Cuyacocha, Chuyayaku, Pindoyaku, Pumayaku, and Wiririma to the east, near of wells identified and marked by oil companies in the 1970s.

This process of space appropriation stood out in 1992, when the Ecuadorian State legalized 251,503 hectares for the Sápara people under the name of "Zaparo traditional settlement title," which provided the first geocultural and geopolitical look at their ancestral territory, now divided between a space recognized and legalized by the State and an unrecognized space of 124,526 hectares (Fig. 12.2). At the same time, the space with no recognition was divided in two: the Kichwa ancestral territory and the Security strip (Table 12.1).

The first is a space of 54,000 ha located within what the State recognized in 1992 as the "Ancestral territory of the Kichwa population of Sarayaku" and for that reason, this space is subject to territorial conflicts between the Kichwa and Sápara people. For its part, the Security strip is a space of 70,000 ha, initially uninhabited, located in what the State established as a safety zone after the 1941 border conflict with Peru. It was legalized and merged in 2011 under the "Zaparo traditional settlement title," however, since the foundation of Wiririma, Garza Yaku, and Nuevo Amazonas, it was a space controlled by people from the Achuar culture. This is because the surface of the Sápara territory is mainly distributed between two parishes (the smallest administrative division of the Ecuadorian state):

(1) Montalvo, created in 1911, during the rubber fever in the Amazon, where the Achuar, Shiwiar, Andoa, and Sápara territories converge; it has always been controlled by the Andoa and Sápara people.

(2) Río Tigre, created in 1921, at the end of the rubber exploitation era, encompasses part of the Shiwiar, Sápara, Kichwa, and Achuar territories (Fig. 12.3). The

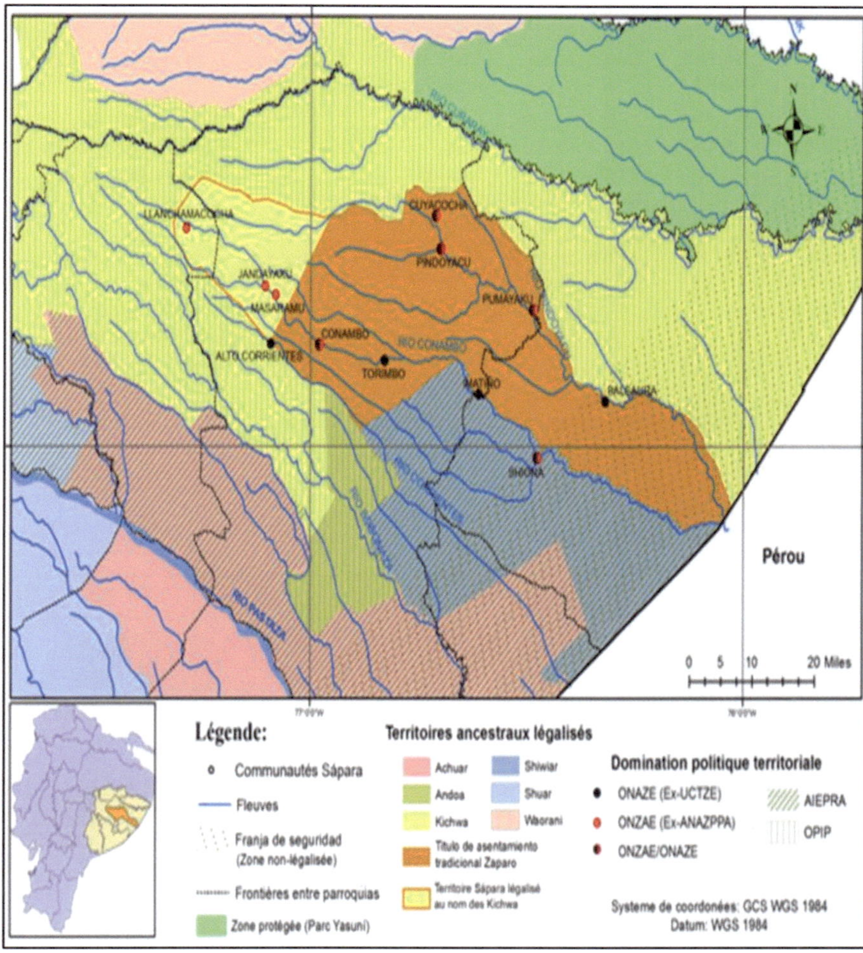

Fig. 12.2 Geopolitical situation of the Sápara 1990–2000 (*Source* Adopted from Mejía, 2014)

representatives of this last territory, find themselves excluded from important positions in the new political organization the Sápara Nation of Ecuador (NASE), founded their own communities inside the former Security Strip, taking power by becoming the headquarters of the Río Tigre Parish (Mejía, 2014).

A Parrish gathers communities without taking into account their cultural, ethnic, political, and territorial affiliation:

> In my conception, the Sápara block should not be defined in political-administrative forms of the state. The main reason is that a people cannot be commanded by another nationality. That in cultural and identity terms is monumental […] There has always been a rivalry between the Shuar and Achuar, Huaorani and Tagaeri-Taromenane nationalities; the Sáparas were so warlike that they sometimes killed each other just to say hello. There are rivalries, and the

Table 12.1 Organization of the ancestral Sápara Territory

Social set	Surface (ha)	Localization	State of legalization	Dominant ethnic ancestry	Communities
Ancestral Kichwa territory	50.000	At the west of the ancestral territory	Not legalized, still in dispute	Zápara	Llanchamacocha, Ripano, Jandiayaku, Masaramu and Cuyacocha
Zaparo traditional settlement	251.000	At the center of the territory	Legalized in 1992	Zápara Kichwa Achuar Mestizo	Ayamu, NimaMuricha, Conambo, Torimbo, Imatiño, Suraka, Alto Corrientes, Lupuna, Pumayaku, Pindoyaku and Chuyayaku
Security strip	70.500	At the east of the territory	Legalized in 2011	Achuar	Shiona, Espejo, Nuevo Amazonas, Garsayacu, Wiririma and Balsaura

Source Mejía (2014)

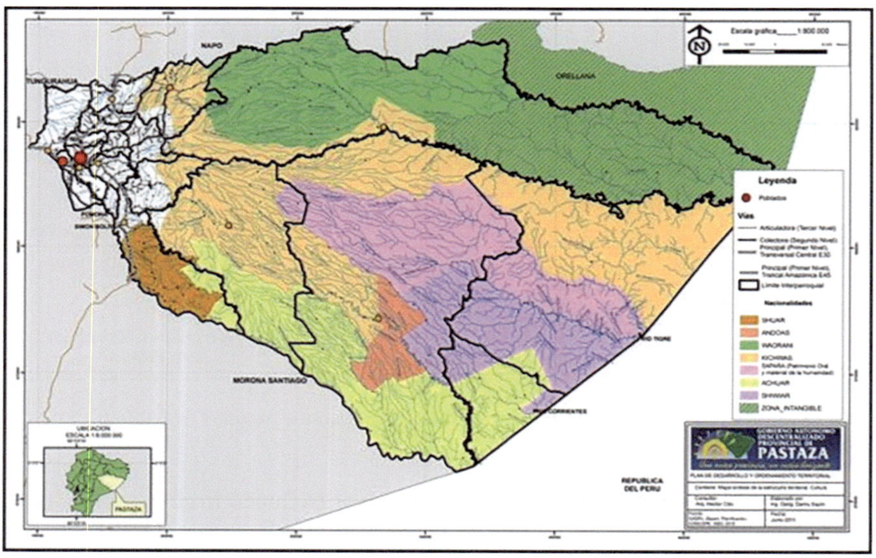

Fig. 12.3 Parochial-indigenous territories overlap (*Source* Development Plan and Territorial Planning of the Pastaza Province till 2025)

territorial ways have always been moments for war, territorial power struggles. So far, that allowed it to happen. The constitution says that the territories that had ancestral settlements must be recovered by nationality, which is the most difficult (Excerpt from an interview with Carlos Duche, Director of the Ministry of Culture, Pastaza Province. Puyo, August the 5th, 2010).

The parish is constituted inside a traditional territory as an important instrument of political power. While communities and political organizations represent the unity and the first form of an independent indigenous organization, a parish is the opposite, provoking conflicts inside and outside a territory and threatening the autonomy and boundaries of traditional or ancestral territories.

12.5 The Limits of the Territory

Although the Sápara Nation of Ecuador represents the interests of the inhabitants outside the territory, in the jungle there are no demarcated borders or a specific type of landscape that distinguishes the Sápara territory from that of its neighbors, so the rivers are the only visible borders and the communities the only guarantee of their existence and respect. Since the communities are separated from each other by hours or days of walking or canoeing, it can be argued that there is an inhabited and an uninhabited space in the Sápara ancestral territory. The latter is represented by the depth of the Amazon jungle, which, although it has no human presence, is not truly uninhabited as it is also inhabited by non-humans, that is, animals, plants, and spirits, which is why it is considered by the Sápara as a biological and spiritual reserve and a source of renewal (Fig. 12.4).

In turn, inhabited spaces consist of a network of communities understood as spatial units delimited by the practices of those who inhabit them. According to

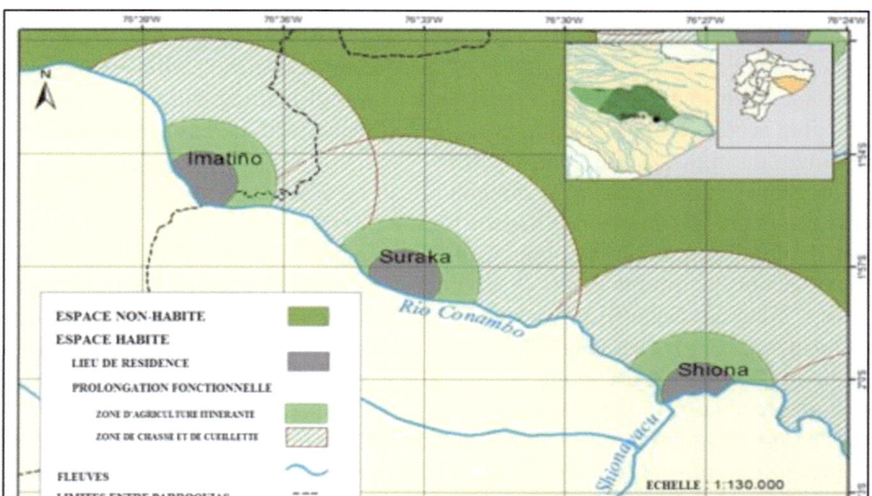

Fig. 12.4 Inhabited and non-inhabited space of the Sápara territory (*Source* Adopted from Mejía, 2014)

Ruby (2003), practice is a human action that fits into an environment composed of other practices, transforming it into an inhabited space. For Lazzarotti (2006, p. 42), it is a geographical practice, which means that it can be considered in terms of its localization and orientation to the place where it is performed: "It is a geographically oriented action." It is possible to state that it is the configuration of a territorial network, as Haesbaert (2004) proposed, in which there are social, cultural, political, economic and environmental dimensions, and possibilities of overlapping scalar, from local to regional spheres. According to Santos (1996), it is also possible to add the technique in its relationship with social practices, as the Sápara world is marked by an intrinsic link between all the dimensions in which Western society divides the world since it is a cosmogonic universe of overlapping vertices. Santos's discussion about space as a system of objects and actions helps to understand space in its relationship with the practices in the life of the community.

Consequently, inhabiting the space far from simply meaning the fact of residing, means the way in which a society develops, performs, or expresses mostly in acts (Stock, 2004), then in practices. This way, the Sápara ancestral territory is established as an appropriate territory, where, as Brunet (1991) points out, appropriate is read in two senses: as a property of the Sápara people and as a space that has the appropriate characteristics for a particular way of inhabiting it, adapting to it or adapting it to that way of life.

With the passage from a semi-nomadic to a sedentary life, it can be observed that the practices of its inhabitants are associated with various places: houses, gardens, schools, airstrips, telecommunications radios, farms, rivers, lagoons, hunting and gathering circuits, as well as tambos. The "tambo" is a Kichwa mot that means auberge, designating a kind of chasse lodge that the families of the communities of certain indigenous people of the high Amazon possess. Located a few hours from the community, the tambo are places of "secondary residence", with small houses or simple cabins and gardens. They are extensions of hunting and fishing grounds providing access to more distant, unfrequented regions. In the past, they were used as refuges during wars (Descola, 1993).

Consequently, the area of a community can be calculated from the location of these places and the trajectories that the inhabitants take to reach them. Thus, we can consider that a community occupies a circular or more or less semicircular extension if it is established on the banks of a border river, with a radius of 7 km, of which the first 3 km are destined for more or less intensive use, and the following 4 km, intended for extensive use (Fig. 12.5). From each point of the centrality of the houses, a main range of action is configured, which is why the living space and its territorial network are mainly circular in perspective.

The first 1.5 km is the residence or community center (Fig. 12.6), determined by the presence of private structures such as houses and gardens, customary structures, and places such as the airstrip, the school, and the communal house; on the other hand, it is determined by the lagoons and rivers. The structures and common places represent the space for gathering, encounter and exchange between families and between the members of these families with other inhabitants or individuals outside the community.

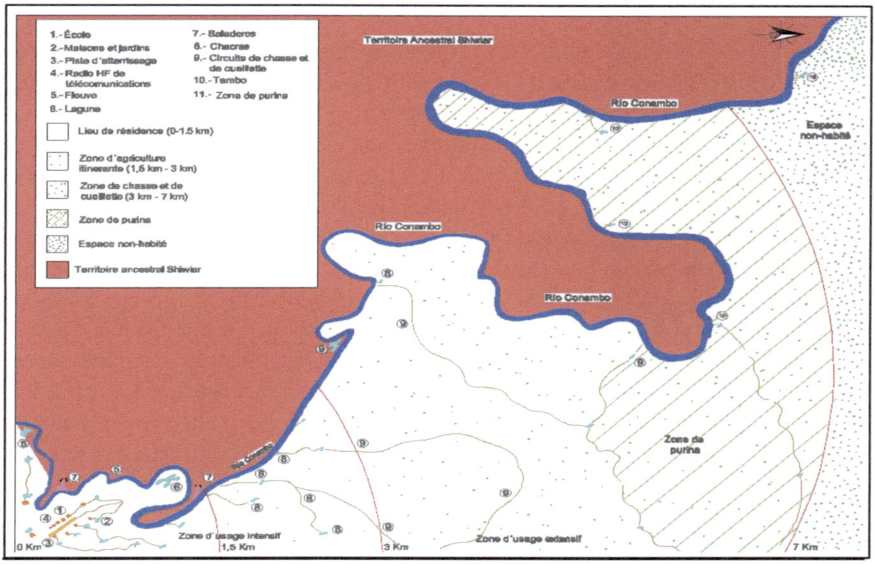

Fig. 12.5 Area representation of a Sápara community (*Source* Adopted from Mejía, 2014)

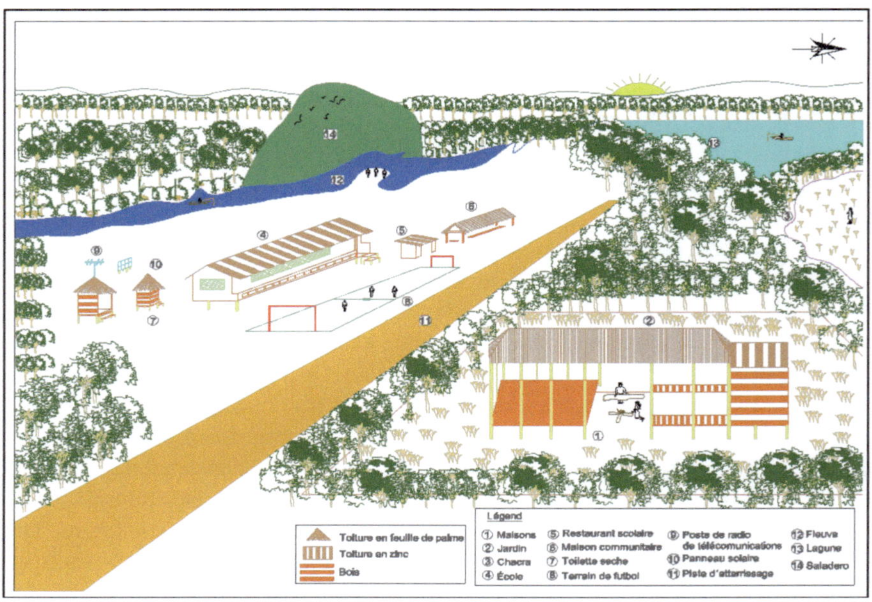

Fig. 12.6 Representation of the place of residence in a Sápara community (*Source* Adopted from Mejía, 2014)

The space outside the home can be considered as its functional extension (Stock, 2004) and is divided into two levels. The first is an area where the inhabitants mainly practice mainly subsistence itinerant agriculture, extending from 1.5 km to the imaginary edge of the first 3 km (itinerant agriculture zone). The second level is the extension of the first, deep into the Amazon jungle, and is where the Sápara mainly hunt and gather (hunting and gathering area). In this context, the maximum extension of the community, where the boundaries of the inhabited space are defined, is only the sum of the collective or individual hunting and gathering routes. The Purina sites and tambos may be located outside this 7 km radius, but in most cases, they are not in the deep jungle, but on the banks of the same rivers where the communities are located. The Purina is a term from the Kichwa language that means "to migrate," "to move," or "to displace" and is one of the main ancestral practices of some Indigenous peoples in the upper Amazon. The Purina was a journey that encompassed the entire ancestral territory and followed the cycles, seasons, quantity, quality, and distribution of fauna and flora and could last up to two years before returning to the starting point. It allowed us to supplement food needs and to transmit and renew knowledge and beliefs from one generation to the next.

Regardless of the above, under traditional living conditions, a community is a place without human presence for much of the year unless there is a school. Although the oil industry has led directly and indirectly to the establishment of communities, it was the school that brought the change from a semi-nomadic way of life to a settled way of life. Considering what it takes to have a school in a community—human capital or physical effort, the difficulty of finding a teacher, and funding, when residents invest more energy to equip and maintain their communities—airstrips, community houses, dry toilets, teacher's house- the desire to leave decreases, even as forest resources have diminished, and soils become infertile (Rival, 2000).

In fact, the sedentary community life has created a new configuration of space, a kind of transposition of the traditional model of displacement that puts pressure on the resources of the only place where it is now practiced. This idea is illustrated by the "Purina." In communities with no school, Purina is still recreated according to tradition. This situation was corroborated on the night of September 17, 2010, when Nina Muricha arrived in the Sápara community, where no residents were to be found since they had all left for the "Purina" (Mejía, 2014). On the contrary, in communities where life is dominated by school, there is no time left to make long trips. Although young people accompany their parents on trips to the charges, hunting, fishing, and gathering sites, the Purina only takes place during school vacations, making it difficult to recreate and transmit knowledge and beliefs to ensure physical and cultural survival in the Amazon jungle, which can have serious consequences for their future. This includes, for example, cutting down the closest forest to the community to cultivate the land. Over time, people have to look further and further afield for the kind of soil they need for their new farms. This means that they are less diverse and yield little profit because they cannot constantly move to maintain them (Bilhaut, 2007).

12.6 Final Considerations: Origin, and Foundations of Tourism in Amazonia

Given the Sápara's living conditions, especially their distance from urban centers and the abandonment from the state, there are not many sustainable alternatives to stop the "perverse effects" of sedentary life in the communities. To ensure their physical and cultural survival, the Sápara must decide whether to allow the advance of oil companies or other extractive industries, develop agriculture and livestock, and/or try to promote tourism. While the first two options offer some opportunities to find employment or gain economic benefits, they also mean the destruction of the forest through pollution, especially of water, and the deforestation of large areas for the construction of roads for the oil companies and the development of monocultures. As in other regions, it is possible that with the construction of oil roads comes settlements and, inevitably, the expansion of agricultural boundaries. On the other hand, without roads, the development of agriculture and/or livestock is impossible due to the cost of transporting products by air to market.

In this sense, the Sápara, like other indigenous peoples of the Ecuadorian Amazon, have decided to receive tourists in their ancestral territories and benefit from the income that this activity brings. However, they also see tourism as an opportunity to make their culture and territory known and recognized, a kind of "renaissance" in a complex present in which the presence of tourists represents a kind of intangible bulwark that stops the advance of extractive industries and agrarian fronts into their habitats. Paradoxically, the conditions that allow the arrival of tourism in this region of the country are still linked to the development of the oil industry. This situation is explained by the physical accessibility of the Amazon through the landing strips and oil roads, which at the same time cause a decrease in access to the natural resources on which survival in the jungle depends.

The territories these people have always inhabited have shrunk due to deforestation, water pollution, and progressive settlement in those cases where oil production has already begun. In areas that have not yet developed, the oil industry has brought about a lifestyle evolution that consists primarily of continuing to develop the traditional socio-cultural practices that once prevailed in a single, small, and therefore limited area of the community. If we understand that the oil industry meant the prohibition or deprivation of sufficient access to resources to sustain life in the jungle, then we can only assume that indigenous peoples were forced to seek new strategies to adapt and survive. They organized themselves at the territorial, local, and regional levels, forming communities and joining together in an organization that represented them as indigenous peoples at the national and international levels, enabling them to become spatial and political agents of their own development.

References

Andrade, C. (2001). *Kwatupama Sápara = Palabra Sápara: Obra maestra del patrimonio oral e inmaterial de la humanidad (UNESCO).* Puyo, ANAZPPA.

Bilhau,t A. G. (2007). *Le réveil de l'immatériel. La production du patrimoine onirique des Indiens Sápara (Haute Amazonie).* [thesis]. [Paris]: Université Paris X Nanterre; 2007. 395 p (in French).

Brunet, R. (1991). Le territoire dans les turbulences. Montpellier, Reclus coll. Géographiques.

Castillo, M., Félix, J., Mazabanda, C., Melo, M., Moreno de los Ríos, M., Narváez, R., Páez, B., & Usigua, M. (2016). La Cultura Sapara en Peligro ¿El Sueño es Posible? La lucha de un pueblo por su supervivencia frente a la explotación petrolera. Quito, Terra Mater.

Claval, P. (2012). *Géographie culturelle. Une nouvelle approche des sociétés et des milieux.* Paris, Armand Colin.

Costales, P., & Costales, A. (1975). La familiaetno-lingüísticaZápara. *Ethos, 1*, 3–30.

Descola, P. (1993). *La nature domestique. Symbolisme et praxis dans l'écologie des Achuar*. Paris, Édition de la Maison de Sciences de l'Homme.

Duhamel, P. (2003). Les lieux touristiques. In M. Stock (Ed.), *Le Tourisme. Acteurs, lieux et enjeux.* Paris, Belin.

Fontaine, G. (2003). *El precio del petróleo: Conflictos socio-ambientales y gobernabilidad en la región amazónica.* Quito, FLACSO, IFEA.

Gobierno autónoma descentralizado provincial de Pastaza. (2009). Mapasíntesis de la estructura territorial, cultura. Plan de desarrollo y Ordenamiento Territorial de la provincia de Pastaza al año 2025.

Haesbaert, R. (2004). *O mito da desterritorialização. Do "fim dos territórios" à Multiterritorialidade.* Rio de Janeiro, Bertrand Brasil.

Lazzarotti, O. (2006). *Habiter. La condition Géographique.* Paris, Belin.

Mejía, V. (2004) Les problématiques et les enjeux de la mise en tourisme du patrimoine culturel immatériel: l'exemple du peuple indigène Sápara d'Équateur. [thesis]. [Angers]: Universidad de Angers; 2014. 350 p (in French).

Müller, A. (2011). Consecuencias del contacto de lenguas en el caso de las nacionalidades Shuar y Sápara del Ecuador. [dissertation]. [Wien]:Universität Wien. 2011. 204 p (in Spanish).

Pérez, S. (2005). Les peuples indigènes et l'éducation bilingue interculturelle. Le cas de l'Équateur. In: L. Prudent, F. Tupin, & S. Wharton (Ed.), *Du plurilinguisme à l'école. Vers une gestion coordonnée de langues en contextes éducatifs sensibles.* Berne, Peter Lang.

Santos M. (1996). *A natureza do espaço. Técnica e tempo.* Razão e emoção. São Paulo, Hucitec.

Ribeiro, G. (2017). *Fernand Braudel, Geohistória e longa duração. Críticas e virtudes de um projeto historiográfico.* São Paulo, Annablume.

Rival, L. (2000). La escolarización formal y la producción de ciudadanos modernos en la Amazonia ecuatoriana. In A. Guerrero (Ed.), *Etnicidades.* Quito, FLACSO.

Rodríguez, S. (2007). Estudio Socio-Histórico de la Nacionalidad Sápara de Pastaza – Ecuador "NASAPE" (No publicado).

Ruby, Ch. (2003). Pratique. In J. Lévy, Lussault (Ed.), *Dictionnaire de la géographie et de l'espace des sociétés.* Paris, Belin.

Stock Mathis. (2004). L'habiter comme pratique des lieux géographiques. EspacesTemps.net. https://www.espacestemps.net/en/articles/habiter-comme-pratique-des-lieux-geographiques-en/

Trujillo, P. (2001). *Salvajes, Civilizados y Civilizadores. La Amazonia ecuatoriana: El espacio de las ilusiones.* Quito, Abya-Yala.

Viatori, M. (2005). The language of "authenticity": shifting constructions of Zápara identity, the politics of indigenous representation, and the state in Amazonian Ecuador. [thesis]. [California]: University of California.

Wladimir Mejía Ayala is an associate professor at the Pedagogical and Technological University of Colombia. Agroforestry engineer and PhD in Geography from the University of Angers, France, he is interested in the relationship between tourism and intangible heritage. Currently, he is the academic coordinator of the Program of Postgraduate Studies in Geography EPG (Uptc-IGAC agreement) and editor of the magazine PerspectivaGeográfica.

Júlio César Suzuki is an associate professor at the University of São Paulo in the Department of Geography and Doctorate Program in Latin American Integration. Graduated in Geography from the Federal University of Mato Grosso and a Master and a PhD in Human Geography from the University of São Paulo. He was director of 31 master's thesis and 32 doctoral thesis.

Chapter 13
Territoriality and Co-construction of Territorial Involvement: Action Research and Popular Geopolitics

Marcos Aurelio Saquet

Abstract This chapter argues in favor of geopolitics and popular geography, either root or native, with a strong territorial basis during the participatory-action-research. Territory and territoriality are, in their natural, social, and cosmological singularities, fundamental for the co-construction of the root involvement, especially for the popular classes. In this conception that we consider decolonial and counter-hegemonic, territoriality contains inequalities and differences, identities and synergies in the geopolitics of environmental and cultural conservation, co-producing knowledge at the university-territory interface, and popular science-knowledge. Thus, initially, we discuss the concept of territory in the historical–critical and reticular conception; then, we relate the territory to the decolonial debate and to popular science and participatory-action-research and dialectics oriented to integrate subjects and knowledge in praxis of sustainable territorial involvement.

Keyword Territory · Cooperation · Popular sciences and knowledge

13.1 Introduction

Over the past 25 years, we have been working on teaching, research, and university extension, seeking to contribute to the creation of a Geography conception focused on cooperation with the subjects of each participatory-action-research project, that is, on the dialogical, ecological, cultural, and participatory territorial development (Dansero, 2008; Saquet 2015 [2011], 2013, 2014, 2016a, 2016b, 2021 [2017], 2018, 2020, 2021, 2022; Saquet, et al. 2012; Scoppetta, 2009). This is our theoretical-methodological and political choice, valuing autonomy in decision-making, the preservation and conservation of nature, identity, popular knowledge, reciprocity, anchorage, and territorial bonds (links), in a movement contrary to the expanded reproduction of capital and the bourgeois State.

M. A. Saquet (✉)
State University of Western Paraná, Cascavel, Brasil
e-mail: saquetmarcos@hotmail.com

M. Mishra et al. (eds.), *Contemporary Geopolitical Alignment in the Global South*,
https://doi.org/10.1007/978-3-031-81977-3_13

Initially, we corroborate Dussel (1995 [1979]) when he states that philosophy must be open to the people, listen to them, and interpret them, as we understand it, with the people in order to, in a participatory and creative way, produce the popular class revolution toward a conscious direction, through a popular philosophy, creating historical commitments based on everyday experience (Gramsci, 1975 [1929–1932]).

This is a process that (im)materializes in class and "place consciousness," in the praxis of social participation, struggle, resistance, decolonization, and counter-hegemony, in which inhabitants feel more alive, important, acknowledged, valued, belonging to the place, to the street, the condominium, the neighborhood, the rural community, the municipalities, to the short networks (circuits), the hydrographic region, the State, the river, in short, to the territory (Saquet, 2018, 2021 [2017], 2021, 2022).

The decolonial and counter-hegemonic geographic knowledge, therefore, as we have been discussing over time (Saquet, 2013, 2021 [2017], 2018, 2021, 2022), must arise from the learnings produced from the denial of the colonizing and oppressive territorial-capitalist order, or from a "geography of absences" that recognizes and values those who become historically and geographically "invisible", generating other theories and methodologies to produce a "feeling-thinking" (*sentipensante*) geography, created with emotion and affection (Quintero 2011; Mansilla et al. 2019). That is why we defend a popular geography, either root or native, with a strong territorial basis during the participatory-action-research.

Hence, knowledge and science, overall, are generated in a process of understanding and socialization of reality, valuing cultural roots, the environment, and people; science has its own roots, and the university reciprocally acts with the people, generating solutions in the vanguard of knowledge in a participatory and flexible perspective that is locally and regionally contextualized, a unique situation called by Fals (2012b [2003]) as "contextual endogenesis".

To socialize our conception of popular, territorial, and decolonial science, we organized this text in two complementary parts, in addition to the Introduction and Final Considerations: we present the option of approaching the territory with a popular and decolonial geography, followed by the description of the international context of discovery and valorization of the territory in different areas of knowledge, considering the participatory-action-research as a territorial praxis of cooperation and struggle in favor of a more just and ecological society. This is a geopolitics and popular geography for the co-construction of the involvement, especially for the popular classes, conception oriented to integrate subjects and knowledge in the praxis of sustainable territorial.

The territory as a fundamental concept and can be summarized as follows: (i) It is appropriate, dominated, and has a political and economic content involving points, networks, and grids (Indovina & Calabi, 1974) that are also affected at cultural and environmental levels, simultaneously, trans-temporally, and transversally. (ii) It is produced at different scalar levels (Magnaghi, 1976; Dematteis, 1985; Raffestin, 1993 [1980]; Saquet 2020 [2007], 2015 [2011], 2021), with a relational content linked to reticular geography (Dematteis 1964, 1985, 1995, 2001; Turco, 1988, 2010) or to a complex cross-scale and multiscale territorial frame (plot) (Camagni & Salone,

1993; Dematteis, 1985, 2001) or, still, to trans-territorial relations (Camagni, 1993, 1997; Camagni & Salone, 1993; Rullani 2009).

From this synthesis that we consider current and quite complete, we are working with the territory from a theoretical-practical and popular perspective. The territory contains and means, firstly, a space of (in)formation, mobilization, resistance, struggle, and confrontation with the great agents of capital and the bourgeois State, in addition to being the object of studies and conceptual guidance in the scientific and popular analysis and interpretation (Saquet 2015 [2011], 2013, 2014, 2021 [2017], 2018, 2021, 2022).

For us, who work with subjects of the people, this concept is fundamental and needs to be complemented with the notion of "transversalities" (Dematteis, 1995; Brunetta, 1997; Deda, 1997; Rullani et al. 2000; Belliggiano & De Rubertis, 2012) and, by forming different networks and grids, they are at the base of each territory (Gottmann, 1973; Indovina & Calabi, 1974; Magnaghi, 1976; Raffestin, 1993 [1980], 1987; Dematteis, 1985, 1994, 1995, 2001; Rullani, 1997), playing a contradictory role of connection and mobility and, at the same time, rooting and territorial bonds.

The interconnection and transversality between university and territory, science and popular knowledge, and society and nature, are fundamental for co-construction of territorial involvement. Therefore, it is essential to recognize that territory and territoriality are always linked: territorialities are constant production—as conditioning and resulting of the life in society and nature—within a historical process centered on power relations, networks, and cultural identities (Raffestin, 1977, 1993 [1980], 2005; Camagni, 1990, 1993; Camagni & Salone, 1993; Dematteis & Governa, 2003; Saquet, 2003 [2001], 2020 [2007], 2009, 2015 [2011], 2016a, 2016b, 2021 [2017], 2021). Territories are (im)material, simultaneously concrete and abstract, objective and subjective, spatial and temporal, and social-natural.

Territoriality corresponds to the unity of social relations, of otherness and exteriority; it is complex and produced in the society-nature relationship through mediators, endowed with changes and permanence; thus, it is at the center of territorial production that has, in its core, "invariable factors" (grids, nodes, and networks) that historically take on different contents and practices aiming at maintaining or achieving the maximum possible autonomy. Territories invariably form grids, nodes, and networks that are daily experienced and full of power (Raffestin, 1977, 1986, 1987). These—"multilateral"—relationships/territorialities take place in the "bio-social" and space–time interface and, as they correspond to power, they are symmetric and asymmetric, permanent and temporary (*plusieursterritorialités*) (Raffestin, 1977).

The historical process is geographical, and the latter is also historical, considering that time and space are inseparable, in a wide and complex movement of deterritorialization and reterritorialization: in deterritorialization, there is a loss of the initially appropriate and constructed territory, the suppression of limits, borders, as stated by Raffestin (1984); and, in the reterritorialization, the reproduction of elements of the previous territory (immediately or not) occurs, at least, considering some of its characteristics (Saquet, 2003 [2001]. Therefore, there is complexity and heterogeneity within/between territories, internally and intermediately, substantiated from the nature-society and time–space relationships, an amalgam that makes our work of research and territorial action very difficult.

Nevertheless, we have been making an effort to simultaneously work on teaching, research, and cooperation (in the countryside, in the city, and in the countryside-city relationship). Cooperation is understood as a process of dialogical construction of ideas, planning, research, and development actions aimed at improving the living conditions of urban and rural people, preserving nature, recovering degraded and polluted environments, valuing the know-how, popular participation, debate, autonomy in decision-making as well as enhancing natural identities and heritage.

This means that thought and action must simultaneously occur, in a continuous movement to create better living conditions through cooperative and solidary forms (Coraggio, 2004 [2000]). Behaviors, such as solidarity, community cooperation, and nature preservation, must be effectively identified, understood, represented, valued, and recreated on a daily basis, paraphrasing Kropotkin (1982 [1885]), with more freedom and independence toward the intellectual and cultural development.

More independence and freedom to think and create, to co-produce science for and with the popular classes, in geopolitics in favor of a more just and ecological society. Science, whether it is natural or social, aims at objectivity through experimentation and reason and is permeated by rules and logic acceptable to specialized scientists (Feyerabend 2010 [1987]). These conceptions can be broadened, considering that science is also theoretically and methodologically produced, as well as by popular knowledge. It is a movement that necessarily permeates the identification, observation, description, tabulation, systematization, analysis, and interpretation of the past, the present, and the future, the latter is also necessarily contained in the participating research and participatory-action process.

Mannheim (1952) understands there is the "theoretical philosophy," which is rationalist, of Enlightenment origin, focused on methodical, scientific, and universalizing knowledge and, at the same time, the irrationalism, aimed at the understanding of the singular, in a subjective and cultural way, apprehending the contents of the forms. The specialized, disciplined, methodical, and rational science that we know and reproduce is "a tradition," among other existing ones that are also important (Feyerabend 2010 [1987]). We must recognize, notably, the importance of reason and rationalism in certain situations; however, there are other ways to critically perceive and understand the world we live in, creating sciences and solutions, especially for the people.

Customs, beliefs, and cosmologies can be useful for certain societies, in certain spaces and times, but not for others. Culture and science are different; however, they are not exclusive. We can all invent, create, and improve; these are characteristics immanent in our minds of men (and women): hence the clock, the machines, the beers, the energies, the architectures, etc., (im)material, theoretical-practical, rational-irrational.

Thus, we may infer there are practical-theoretical experiences, as well as the coexistence of both in everyday life: (a)theoretical elements and processes are mixed and both may or may not be rational, that is, we recognize the existence of rational and irrational (a)theoretical knowledge. Our conception is that there are philosophies and at least one philosophy of/in everyday life (linked to the time of the present/now—without disconnecting from the past, future, and the space of the here—without disconnecting from trans-multiscalar)—(ir)rational practical-(a)theoretical or (im)material (concerning forms and meanings), discontinuous in time and space, both political-environmental and cultural-spiritual. Thus, for us, this is the popular philosophy that comprises in itself a movement composed of practices, and practices in trans-territorial and trans-temporal movements, substantiating themselves in science and popular education. The liberation praxis, therefore, must be, (im)materially speaking, a philosophy of everyday life and "popular science" (Fals 1981, 2012 [1997]), in a movement of struggle and confrontation developed inside and outside schools.

Then we can produce, for and/or with the people, a science of our own, contrary to what is currently put into practice. Thus, we figure that we will be more motivated and committed to the care we must exercise with the territory, understood as our home and place of biological and social reproduction. Apprehending perceptions, attitudes, gestures, local knowledge, opinions, smiles, sadness, etc., is part of the (im)materiality of the (a)theoretical knowledge. To do so, the intellectual must get rid of an academicist and elitist ideology.

13.2 The Discovery of Territory and Its Relationship with Decoloniality and Popular Geopolitics

As we have shown in Saquet (2018), Milton Santos(1994)had specifically warned us of the importance of the "return of the territory", contiguously formed and within a network, in a dialectic that is part of human life. Like domination, resistance is at the place, a fact that makes it essential to understand the territory, as it reaffirms itself through the place at different scales (Santos, 1994).

Place and territory, therefore, are inseparable, although they are different as social and spatial concepts and processes. The aforementioned "return of the territory" happens, from our perspective, from the 1960s to the 1970s, according to the singularities of the different countries, as detailed in Saquet (2020 [2007], 2015 [2011], 2018, 2021, 2022), directly influencing their insertion as theme, concept, analytical category, and social-natural process into different areas of knowledge.

The territory became more diffuse from the 1970s to the 1980s, with the so-called crisis of Fordism and the discovery of Marshallian Industrial Districts, announcing the post-Fordism or flexible accumulation, through localized, contextualized, and territorially articulated processes, characterizing a movement of "reemergence of the territory" (Rullani, 2003), also called by the same economist as "rediscovery of the territory" (Rullani, 2003, 2005), similar to Dematteis (1981 [1979]): "discovery of the territory."

With this conceptual-analytical sense and the spatiality of social processes, the territory is a mediator in the transition from Fordism to post-Fordism, introducing values and competitive advantages into the market (Rullani, 2003). This became the key point in the creation of the heterodox paradigm in the 1970s, interpreting the development problem of what was considered as the "new holy trinity" of the regional economy, composed of technological, organizational, and territorial elements, spatially concentrated and dispersed, transcending production in regional contexts (Storper, 2003).

And, regarding this historical process of revaluation of the territory, it is noteworthy that Dematteis (1981 [1979]) identified this movement already in the 1970s, analyzing it based on the class struggle that occurred in Italy in 1968–1969 (also occurred in other countries), more precisely within the praxis of the struggle of Italian workers. The territory have meant and still means, in different spatial and temporal situations, a local condition of the class struggle, an aspect that we highlighted with Saquet (2020 [2007], 2016a, 2021 [2017]).

In order to know the territorial singularities and to value them as a place of life with its own identity, it is necessary to invest in the scopes of debate, participation, and local planning (Quaini, 2004), understanding the places to create the future (Magnaghi, 2000), with the (im)material meaning of "return to the territory" (Becattini, 2009; Magnaghi, 2015), as we shall further detail.

The discovery of the territory precisely occurs when the distinctive identity of the territories comes into play, inducing feelings of belonging and sharing, together with the very management of conflicts based on the definition of common interests and objectives. Thus, development is understood as a localized process, with specific characteristics, as stated by Rullani (2003, 2005). The theoretical framework that stands out is that of the "territorial analysis paradigm" (Bagnasco, 1978), identifying limits, advances, contradictions, inequalities, differences, cooperation, degradation, preservation, identities, etc., to directly support our participation in such dialogical, participatory, ecological, and cultural processes.

Hence, this territorial analysis is necessarily performed with the "reticular paradigm" (Camagni, 1993; Camagni & Salone, 1993), interpreting the different types of networks (short, intermediate, and long, concrete and virtual, ephemeral and permanent), seeking to characterize heterocentric and self-centered territorializations (Turco, 1988) and leveraging the cooperation and solidarity networks locally anchored and in ecological and cultural processes (Saquet, 2021 [2017], 2018, 2021, 2022; Saquet and Alves 2015).

In the conception that we consider decolonial and counter-hegemonic, territoriality contains inequalities and differences, identities and synergies in the geopolitics of environmental and cultural conservation, co-producing knowledge at the university-territory interface, and popular science-knowledge.The networks and territorialities are always present and signify conflicts, contradictions, class struggle (power relations: otherness and exteriority), territorial disputes to "activate territorialities" (Raffestin, 1993 [1980]; Dematteis, 2001; Governa, 2001; Saquet 2020 [2007], 2015 [2011], 2021 [2017], 2018) as an involvement oriented to the cooperation through a transformative praxis (dialogical, critical and reflective) to satisfy people's needs and achieve as much autonomy in decision-making as possible (Freire, 2011 [1996]; Vazquez 1990 [1977]; Dematteis, 1985, 1995, 2001; Magnaghi, 2000; Raffestin, 1993 [1980]).

In this sense, the territorial "anchorage" (*ancrage*)(Pecqueur & Zimmermann, 2002), or *arraigo* (Chávez and Salcido 2014; Sanz, 2014; Harguindeguy, 2014; Bartra, 2014a, 2014b), also has a key role in local bonds and short networks oriented for the territorial involvement. And this problematic of territorial anchorage is not as recent nor unknown, as we have already pinpointed. Becattini (2000 [1989]) clearly understands it based on the importance of community life, the mutual influence between small and medium-sized companies, the existence of local networks of specialized transactions, and differentiating and valuing the specificities of each community and the local bonds that exist there. Bagnasco (1977, 1988) also understands that in territorial development there are productive connections in the territory, local bonds, traditions, relationships of trust, recognition, and identity between companies dependent on each other; specific cultural ways of producing, continuing to incorporate technical and technological innovations.

Thus, together with anchorage, we must highlight the importance of *circuits courts,* understood as geographical and institutional proximity (Chaffotte and Chiffoleau 2007; Chávez and Salcido 2014; Sanz, 2014) or *filierecorte* (Belliggiano & De Rubertis, 2012; Dansero and Puttilli 2014; Forno & Maurano, 2016; Corsi et al. 2018) or as short-circuiting (Renting et al. 2003).

This process can take place in a continuous movement of quantitative and qualitative overcoming, which can occur—without disregarding the crucial importance of the state—through self-organization and self-management (Dematteis, 1985, 1995, 2001; Magnaghi, 2000). Obviously, we cannot forgo a State intensely acting for its social attributions and, at the same time, directly contributing, consistently and permanently, to the provision and/or qualification of living conditions of all people, toward a more just and ecological society.

The territory brings together, in its place, community, institutions, economy, and nature, within a process of personal and social interactions, with intelligence and rooting; it corresponds to the history of human beings who inhabit places, to social practice, and to cumulative knowledge (Rullani, 2003, 2005). Each place must be specifically thought of in a network of territorially connected places (Rullani, 1997).

That is why we have thought about and put into practice the intrinsic "territory-place" relationship (Saquet 2020 [2007], 2015 [2011], 2014, 2018, 2021 [2017]), understood as space–time where people live, feel, perceive, breathe, aspire, suffer, interact (social and social-naturally), recognize, struggle, resist, degrade, preserve. Where people are extorted, exploited, subordinated, and can reorder power relations to live in a more symmetrical way (Raffestin, 1977, 1993 [1980], 2005; Magnaghi, 1976, 2000, 2003, 2015; Dematteis, 1985, 1995, 2001; Camagni, 1993, 1997; Quaini, 2004, 2006, 2010; Rullani, 1997, 2009; Rullani, Micelli and Di Maria 2000).

At the same time, places have "territorial qualities": territorialization and territorialities generate places as one of their historical configurations that must be understood, represented, and explained together with the territory (Turco, 2010). The territory, in turn, contains the "nature of the plurality of places" (Becattini, 2009 [2002]). And this understanding of territory-place is essential for understanding class and place consciousness, without the usual separation between society and nature, time and space, territory and place, university, and territory, influencing the praxis of social and territorial transformation in favor of the people, their needs, aspirations, and desires.

The aforementioned praxis should consider the sense of belonging to a "social class and to a place" (Harvey, 1982; Becattini, 2000 [1989], 2000, 2009, 2015; Magnaghi, 2000, 2015; Quaini, 2010; Becattini & Magnaghi, 2015; Saquet 2015 [2011], 2013, 2014, 2016a, 2021 [2017], 2018) in a movement contrary to the expanded reproduction of capital, to environmental degradation etc.

The "place consciousness" is vital for social identification and participation, for the reproduction and qualification of the territories of coexistence with community relations, which can mean a possible antidote to economic globalization (Becattini and Magnaghi, 2013, 2015), being inspired by and practicing values effectively more human and ecological, aiming at the emancipation of everyone.

The territory, for example, in the Ts'umbal group, composed of active and collective subjects, corresponds to the *lumk'inal*, that is, to the place of daily life, characterized by its histories, cultures, and singularities, establishing power relations among themselves and with subjects from other territories: therefore, the territory is the source of its strength as a social group, to fight for its rights when facing the social transformations taking place in Chiapas—Mexico (Sántiz-Gómez, 2015). This relation of territory-culture unity signifies an identity potentialized as political resistance among the indigenous people, similar to what other authors did, such as Chamorro (1998), when studying the cosmology of the Guarani people recognizing deep connections between human beings and deities, as one of the elements that characterize them as a social group to reproduce themselves as such.

In this sense, it is worth noting that, in *tseltal* (language spoken among Indigenous people of Chiapas, Mexico), the meaning of living well (*lekilkuxinel)* is experienced, according to the Indigenous culture, with its history, in an integrated view of life, considering family, environmental, economic, and cultural aspects: the search for a full life (*lekilkuxlejal*) is directly associated with the present and future of Indigenous families (Sántiz-Gómez & Parra Vázquez 2017).

We believe, therefore, that the formation and/or qualification of class and place consciousness is, at the same time, fundamental to the construction of a "counter-hegemonic paradigm" of research, teaching, and cooperation, considering the meaning addressed by Sánchez (2014), Harguindeguy (2014), Hidalgo and Fernandez (2012), Quijano (2000), Bartra (2014a, 2014b), Fals and Mora-Osejo (2004), Reyes (2009), and Lizárraga and Vacaflores (2014), trying to break free from hegemony of the colonialism perpetuated by Eurocentric paradigms (Quijano, 2000; Fals and Mora-Osejo 2004).

The construction of the "counter-hegemonic paradigm" must necessarily be participatory and dialogical, reflective and creative, decolonial and popular geopolitics, involving and valuing the differences, inequalities, and identities of each place and territory, without disconnecting from the world, though highlighting the close relationships, solidarity, popular culture, anchorage, artisanship, agroecological practices, the specific edaphoclimatic conditions of each ecosystem, etc.

This movement can be understood as a "return to the territory," with the meaning advocated by Becattini (2009), Magnaghi (2015), Becattini and Magnaghi (2015), and Saquet (2018), from the perspective of identifying, understanding, representing, interpreting, valuing, and enhancing the singularities (social, natural and cosmologic) of each place-territory. This process of "return to the territory" requires the explanation, and valuation of anchorage, proximity (short networks), bonds, belonging, self-management, and, at the same time, the centrality of class and place consciousness as a mediating praxis in the daily struggle and in the promotion of territorial (de)involvement with reciprocity, environmental preservation, dialogue, and popular participation. Class and place consciousness conditions the return to the territory, and the territory influences the qualification of class and place consciousness.

This also requires a conception (of popular research and cooperation) that effectively recognizes the society-nature, university-territory, science-popular knowledge, and territory-place relationships, which are affected by our direct and systematic participation in involvement processes in the sense that we briefly describe in this text and detail in other studies (Saquet 2015 [2011], 2013, 2014, 2021 [2017], 2018, 2021; Saquet and Alves, 2015).

13.3 Territorial Trialectics in Participatory-Action-Research

Hernández Cordero (2008) also evidences, in the comparison with the dialectic of Milton Santos, the trialectics used by Edward Soja through a holistic conception of space, centered on the space–time-society relationship. Hence, the subjects historically and socially produce the space, therefore, geographies, histories, and societies, in which they reproduce representations and practices from their daily experiences. Subject and object are different, but human beings are not separated from their objective and subjective existence in time and space.

This trialectics can also be identified in the useful reflections of Calvino (2008 [1972]) on the invisible cities existing in the imagination of Marco Polo and Kublai Khan—coexistence that supposedly occurred in the thirteenth century—through the memory-desire-symbol and/or city-imagination-representation and/or visible-invisible-interpretation, and/or feelings-totality-metaphors relationship.

> Marco Polo describes a bridge, stone by stone.
>
> 'But which is the stone that supports the bridge?' Kublai Khan asks.
>
> 'The bridge is not supported by one stone or another,' Marco answers, 'but by the line of the arch that they form.'
>
> Kublai Khan remains silent, reflecting.
>
> Then he adds: 'Why do you speak to me of the stones? It is only the arch that matters to me.'
>
> Polo answers: 'Without stones there is no arch'
>
> (Calvino, 2008 [1972], p. 79).

We can also mention other authors, such as Bagnasco (1977), which puts into practice a trialectic and trans-scalar approach to the Italian uneven and territorial development, centered on the society-space/region-time relationship and on political-classes-regions companies/institutions. Local societies are formed by multiple elements and economic, political, and cultural processes, being able to characterize groups organized in continuous or discontinuous productive circuits, due to the multiplicity of subjects involved there, relationships, and networks that result in any condition, at the same time, different "territorial articulations", in time and space (Bagnasco (1977, 1988)).

Furthermore, we can revisit the trialectics conceived by Kosik (1976 [1963]), based on the past (animality)-present (sociability)-future (human world) relationship, established through the work process as objectification and unity of humankind with nature. Three-dimensionality also defended by Braudel (1992 [1969], 1981 [1977]), through brief, short, and long times, featuring temporal trialectics, in which history is a dialectic *of* social duration, past and present, which manifests itself in coexisting spaces. Hence, there are rhythms coexisting in space, as civilizations coexist in the "imperious time of the world," (re)creating "polarizations," "depolarizations," and "repolarizations" in the different "world economies," intertwining long, brief, and short times, also simultaneously: historical and simultaneous "multi-secularly" times coexist (Braudel, 1981 [1977]).

From these conceptions, although they are different at the philosophical level, we can deduce that, in all of them, there is a centrality of the time–space-society relationship. Certainly, our representation is reduced in relation to the conceptions of these important researchers, however, it is important to show the unity between the social-spatial–temporal processes, regardless of the territory where we live. We, as men-society-nature, are space and time in continuous movement, we are in a certain space–time transformed daily into territory.

In this way, our life is made up of coexistences of imaginaries, rhythms, trans-scalar articulations, animalities, sociability, stories, etc., as stated by I. Calvino, F. Braudel, K. Kosik and A. Bagnasco, in different areas of knowledge. This coexistence and simultaneity is fundamental to understanding what might be called spatial–temporal-social-natural-cosmological gravitational and quantum polylectic.

The dialectic-trialetic-polylectic relationship is fundamental to understanding the singularities of each territory, ecosystems and microclimates, soils and vegetation, as well as individuals, families, groups, and social classes in a broad process of territorialization-deterritorialization-reterritorialization. In this way, we will be able to understand and represent the changes and permanences, the differences and inequalities, in short, the heterogeneity of each territory and between different territories, identifying networks of cooperation and solidarity (community relations), self-organization and synergies, popular struggles and resistances, anchors and intelligence to the co-construction decolonial and counter-hegemonic.

And, for this decolonial and counter-hegemonic co-construction, rediscovering the "philosophy of liberation" is very important, from the sociology proposed by Orlando FalsBorda, the pedagogy of Paulo Freire, the understanding of praxis as a mediation to understand and transform reality in favor of the other, that is, of the people, with love and solidarity (Dussel, 1995 [1979]). It is a liberating praxis contrary to the dominating one (Dussel, 1995 [1979]) or a revolutionary praxis contrary to the manipulative one (Kosik, 1976 [1963]).

The philosophy of praxis unravels the human beings' self-creation as natural, social, and spiritual/cosmologic beings, within a human being-world, work-existence unity. Therefore, daily life is paramount, such as territorialities and temporalities, as well as the organization of daily, family, and individual life, with repetitions, rhythms, actions, reactions, contradictions, disputes, etc., precisely as subjects in a given space–time-territory relationship.

The advocated historical–critical, (im)material and territorial interpretation cannot occur disconnected from the working action, inside and outside the university: knowledge and action are conditioned, knowledge becomes action and action becomes knowledge in a participative way, in praxis, in the movement of confrontation and resistance, of substantiating popular and territorial science. The latter comprises proletarian, peasantry, *quilombola*, indigenous science, etc., effected, therefore, as con(science), emotion, and reason: class and place consciousness with popular science—and such with class and place consciousness, made of practices, philosophy, and a decolonial and counter-hegemonic movement.

Popular science is also composed of attitudes, procedures, gestures, mimics, choices and decisions, theories and practices, inside and outside universities: practices there are theorized and theories put into practice; systematized and reflected, then, especially in academic spaces, where the new often emerges and that is also, in certain situations, present in the life of the people. In popular life, science and philosophy are not separated, similar to what happens in the art-philosophy relationship. Politics and common sense intertwine, as well as our biological-social-spiritual reproduction.

The everyday present, then, becomes vital as becoming without disconnecting from the past: the present with a future to be debated and created through a possible utopia. Thought and being are identical and constitute a single element, namely, people as a social-natural-spiritual being. Popular and territorial science, therefore, is a movement of simultaneous production of knowledge, thoughts, and actions, participating research and participatory action (also necessarily popular and territorial). Thus, there is also a popular, theoretical-practical thought (Fals 2012a [2003]).

Popular science is at the service of the popular classes and is conceived with them, against the imperialist and oligarchic domination: grassroots groups are subjects and not objects exploitable by the researcher. It is a science of the present and the future, made with the involvement of the researcher, directly participating in the studied process by teaching, learning, and seeking to contribute so that the claim and struggle are continuous and autonomous. To do so, self-knowledge, interdisciplinarity, and political awareness are fundamental (Fals 2012 [1972], 2012 [1981]), a process in which systematization, analysis, reflection, and synthesis remain key, that is, there is no loss of scientificity due to its directly participatory and popular political content.

In our university experience, research is carried out with the subjects, and also the other actions; decisions are made with the subjects of each action-participation-research: I research and work; I work and research. I work during and after research. I research during and after the action. The reflection is constant, as well as the systematizations, discussions, data collections, territorial transformations, etc., in a single process of theoretical-practical praxis: knowledge is transversal/trans-territorial, multidimensional, trans-temporal, and cumulative.

We work to create a praxis of liberation. And, in this form of researching-working, we consider knowledge emanating from several sources, tensions, and consciences: in the political pedagogy of the popular classes, discussions are held, and decisions are made with autonomy before the State and academic professionals, contributing to form active and creative individuals (Fals 2012 [1986]).

The application of techniques and technologies for the representation of geographic information, in a participatory way, can signify an important step for the territorial interpretation of the daily life of peasants, Afro-descendants, and indigenous people as well as their management of common natural goods. Participation must occur from data collection, going through the discussion of methodology to the planning and implementation of actions aimed at the territorial involvement.

Such an approach, in addition to being agile, has a character of intensive and interactive learning, generating, for the studied population: i) knowledge of their territory and political mobilization for self-management; ii) the possibility of analyzing and defining its priorities considering new technologies and social inclusion; and iii) the possibility of planning actions and of evaluating the performed activities. The use of a participatory methodology stimulates the capacity for cooperative planning, debate, school-society integration, and community development.

In popular science, the deification of famous authors is repudiated—as if they were the only ones to think and reflect—in addition to the reverence/obedience to preconceived theoretical schemes, as if they were self-explanatory of the social complexity: it is a science contrary to colonialism. From a "methodology of consensus," we move toward a "methodology of contradiction," with popular commitment, respect for the local society, social insertion/immersion, and without the subject-object dichotomy (Fals 2012 [1972]).

Popular and territorial science can be easily accomplished based on the Indigenous thought, producing knowledge by thinking for oneself and to oneself, as a community anchored in its territoriality, producing its own way to know and learn without leaving the community, inside the community itself, in its core or heart, paving the way of collective knowledge to generate solutions that are also collective. It is an intersubjective and dialogical process between subjects (social and natural) and their integration into the world, understanding it without being detached/disconnected/separated from it, generating an "indigenous science" or "another science" (Quintero, 2011).

The others think and feel, eat and breathe, have wishes and needs, (im)materially revealing themselves through signs and symbols, rituals and gestures and, when attentively, credulously and carefully listening to them, we can deeply understand them, describing and reflecting, reflecting and describing: thinking with the heart and feeling with the mind, this is a serious and urgent need on our part, as educators and researchers, thinking and feeling at the same time. Therefore, putting oneself in the others' place is paramount, as well as understanding their trajectories, stories, memories, narratives (including, in this moment, the so-called unspoken comments), desires, symbols, objects, etc.

The territory, from the perspective of the Mapuche Kimun Indigenous people (Chile), means the foundation of their existence, of their way of being Indigenous, of their feeling-thinking ancestrally constructed to be reproduced as such, resisting and fighting against invaders and threats (Melin et al., 2016). As these authors state, they support a proposal aimed at the decolonization of the being, knowledge, and being at the territory, based on the Mapuche people's own knowledge (practical and reflected, interdimensional), since they understand that the territory is also a "source of your collective memory" reproduced from its most remote ancestors.

Instincts and sensations can emerge from people's lives when instigated to manifest themselves and are key in apprehending the subtleties, the worldviews that cannot be adequately and profoundly understood from a distance. The participatory-action-research methodology must be elastic and versatile, centered on self-learning, without disregarding, obviously, other correlated processes. Within this way of researching and working, conflicts and disputes, identities and differences still consist in potentialities to popularly create and build. The participatory-action-research is, then, a theoretical-practical methodology to produce knowledge linked to the popular classes (Fals, 2012 [1997]).

Of course, secondary data, bibliographic and documentary research are important, together with precise cartographic representations; however, the diversity of research techniques is essential in the pursuit of guaranteeing different "angles" and the possibility of analyses and interpretations, especially qualitative ones. The qualitative meaning is not in the data or in the technique, but in our minds, in the cognitive and reflective processes, in our feelings, in intuition, in the imagination, and creativity for the interpretation based on our objectives, the problematic of studies, our political-ideological option and the expected results.

And how subtle and difficult is to perceive the smile, the brightness of the teeth, the folds of the skin on the face, the movements of the eyes, the intensity of breathing, the murmurs, the gestures, and the silences!

> As I was leaving the courthouse on my way back to the van, I recognized for a moment the smell and color of the summer evening. In the darkness of my mobile prison, I could make out one by one, as if from the depths of my exhaustion, all the familiar sounds of a town I loved and of a certain time of day when I used to feel happy (Camus 2020 [1942], p. 101).

The participatory-action-research methodology is, necessarily, transversal and versatile, quantitative and qualitative, produced with popular political commitment and social immersion: the scripts for questionnaires and interviews are produced with the subjects of each project, the social cartography is paramount, together with life stories, workshops, participatory observations, in short, by daily living together. And, for us, this is a question of the geopolitics of knowledge and, therefore, scientific and popular.

At the same time, human beings produce their representations, impressions, and metaphors as well as their values and behaviors. Human beings always have a certain understanding of reality, as they are "within" the continuous-discontinuous movement of their creation, reproducing themselves as social, natural, and spiritual subjects. Thus, experiencing the place of daily life is essential to know it and to achieve liberation through community political action, configuring specific territorialities in each territory (Quintero, 2011).

Our daily experiences and territorialities are key, because "[...] all social groups have their own geographical knowledge, which arises from the relationship with their territory as a living space and through the exercise of their territoriality" (Mansilla, et al. 2019, p. 152).

Hence the centrality of an adequate methodology to apprehend the complexity, the contradictions, the identities, the differences, the disputes, the conflicts, the daily manifestations, the representations, etc., a movement of the thought that knows more and better the reality, in a hybrid between science and philosophy of the everyday life.

Hence, the mediate knowledge expresses itself and reveals itself in the immediate, apparent knowledge, one is contained in the other, as its visible manifestation, as a phenomenon of everyday reality. Appearance and essence, one is part of the other, in a dialectical unity and totality, in thought (which is also reality) and in practice (which is also reality). Words and gestures demonstrate a form of language, of communication, and, simultaneously, the subjects' thoughts, actions, practices, etc.

We can say that both co-construct and co-creatively, their development, in a contextual and relational, participatory and dialogical, collaborative and community perspective (Grignoli, 2019; Saquet, 2021, 2022). Creation is relational, and interactive, as expected; however, respecting, collaborating, dialoguing, reflecting, and facing, in praxis, the dominators and oppressors together.

13.4 Some Final Words

The perspective which we resume and seek to broaden by dialoguing in the projects we have already carried out, conceiving a historical–critical, reticular, relational conception that effectively recognizes and considers society-nature and space–time relations, in (in)formation, research, and action, in a Geography of territorialities and temporalities aimed at cooperation, at the development with more social justice, environmental recovery, solidarity, participation, and preservation of the territory in its multidimensionality as a heritage of humanity (Saquet, 2015 [2011]).

Conversely, in territorial processes, following this thought-reality correlation, we can understand that, in *multiterritoriality*, different dimensions can be separately experienced; in interterritorial, there is emphasis on at least one dimension of everyday life; in *pluriterritoriality* and *transterritoriality*, although there are different levels of life, people simultaneously live plural relationships, that is, procedural relationships in time (historical) and space (relational/reticular), horizontally and vertically, in which there are coexistences, temporalities, territorialities, and transversal overlaps.

We believe there are different levels of knowledge production and, also, of action-research (with knowledge), and both, together or separately, can be conceived in a horizontal and/or vertical transversality, that is, while the multidisciplinary level can correspond to a more sensitive and immediate level (separate disciplines), the *pluridisciplinary* and interdisciplinary levels can be characterized as intermediaries (with emphases between related disciplines) toward the coexistence of levels that overlap in the reflexive (popular-scientific), solidary, and cooperative mediate knowledge, of creative co-creation, participatory-action-research as decolonial and counter-hegemonic praxis.

We have endeavored to work, over the years, on the theory–practice interface, of (in)formation-research-action, in a praxis of trans-temporal, trans-territorial, trans-disciplinary, subversive, decolonial, and counter-hegemonic popular science, in a science of the past that is, above all, of the present future.

The soul, hence understood as a kind of spirit of the place and territory, therefore of the subject, is a fundamental component of its (im)material constitution, contributing to characterize territories that are also (im)material, produced by subjects who think, feel, eat, breathe, suffer, walk, create, imagine, perceive, etc. (Saquet, 2020 [2007], 2021). Some (a few, it seems) struggle, resist, and face domination, exploitation, extortion, and oppression (in)forming, guiding, clarifying, singing poetry, their art of thinking and knowing, therefore, of teaching and learning, inside and outside the school.

Self-organization and self-projection of the future are paramount, without disregarding, obviously, the importance of public policies that can, although this rarely occurs, value and encourage initiatives of the popular classes in equally popular geopolitics. We know that it is vital to study the past, as we normally do and have done very well in geography; however, perhaps it is even more vital to co-construct the present future for and with our people. In our research-action-reflection perspective, trans-temporality "points" to the future, as predictions are possible and necessary, scientifically feasible, when we break free from intellectual and political-economic colonialism, working with other theories, methods, and methodologies, territorially and popularly created.

Territorial development, as we have already observed in Saquet (2015, 2011, [2017], 2021), can be heterocentric, when it corresponds to networked territories organizations, and self-centered, when it is (im)materialized through networks of cooperation and solidarity territories. Development, then, re-signified as we have been alluding to for several years, means developing and involving, contributing to eliminating oppressive, subordinate, and dominant involvement, to once again involve a field of cooperation and solidarity, of sharing and synergy, co-constructing the territorial involvement, mobilization, struggle, resistance, and confrontation, a process in which there is no social capital, but rather territorial and humanitarian heritage, in which there are tastes and smells, flavors and knowledge. We are and we have been producing heritages every day and, therefore, we need to make them visible, activated, and valued, we must encourage them, listen to them, etc., directly contributing to the liberation of subjects through shared and community, communicative and dialogical processes, with autonomy in decision-making and self-management, overcoming the colonizing and oppressive involvement.

Our geography for, of, and within the praxis is made for social movements of resistance and struggle, based on them and with them, it is at the political level of the root and own territorial involvement, considering, necessarily:

- Our involvement with its struggle causes, with participatory commitment and dialogical respect, activating/motivating each other.
- Continued (in)formation, due to our unfinished development.
- The decolonization of minds and practices is carried out in the territorial, popular, and counter-hegemonic praxis.
- The cultural and political anchorage, as well as the proximity (organizational, philosophical-political, and ecological) preferably carried out by short networks of cooperation and solidarity, valuing and preserving, therefore, the community bonds.
- Mobilization and confrontation aimed at class and place consciousness as well as environmental preservation and/or conservation.

The territorial, natural, and social heritage encompasses areas and networks, subjects and knowledge, techniques and technologies, languages and memories, etc. And it is in it and from it, the territory-heritage, that the peasantry struggles to reproduce itself as a class through family work relationships, diversified production, short networks of commercialization, agroecological and artisanal practices, cooperation and solidarity, confrontation and autonomy in decision-making, (im)material possibilities that we have been experiencing since our family formation centered on a Catholic, charismatic, and leftist culture, with sincerity and spontaneity, self-organization, struggle and confrontation in the face of oppression, corruption, favoritism, interference, private appropriation of the public space, discrimination, environmental degradation, etc.; therefore, there are no formulas and models, because the involvement is unique in each time-territory, it is experienced, lived amid the everyday complexities, temporalities, and territorialities.

Our bodies are trans-temporal and trans-territorial by nature (natural-social-spiritual), therefore, (im)material, and are in a millenary coevolution, so much so that our parents had a long way to go (both in terms of time and space) until we were naturally, socially, and spiritually conceived. The divisions modern *x* post-modern, concrete x abstract, appropriation x domination, society *x* nature, and zone *x* network only take place in our minds, even in the minds of those who consider themselves as post-everything. Hence, considering the extremely serious living situation of our people, we must increasingly co-inhabit and cooperate with the subjects of each project, assuming, in the praxis, our social responsibility as civil servants. And in geography, when conceived as a popular and territorial science, we must understand trans-territorial and trans-temporal life as such, transdisciplinary and transdimensional, coexisting, cooperating, and fighting alongside our people.

Development, then, re-signified as we have been alluding to for several years, means developing and involving, contributing to remove oppressive, subordinate, and dominant involvement, to once again involve a field of cooperation and solidarity, of sharing and synergy, co-constructing the territorial involvement, social participation, respect, mobilization, struggle, resistance, and confrontation, a process in which there is no social capital, but rather territorial and humanitarian heritage, in which there are tastes and smells, flavors and knowledge. We are and we have been producing heritages every day and, therefore, we must make them visible, activated, and valued, we must encourage them, listen to them, etc., directly contributing to the liberation of subjects through shared and community, communicative and dialogical processes, with autonomy in decision-making and self-management, overcoming the colonizing and oppressive involvement.

The territorial, natural, and social heritage encompasses areas and networks, subjects and knowledge, techniques and technologies, languages and memories, etc. And it is in it and from it, the territory-heritage, that the peasantry struggles to reproduce itself as a class through family work relationships, diversified production, short networks of commercialization, agroecological and artisanal practices, cooperation and solidarity, confrontation and autonomy in decision-making, (im)material possibilities that we have been experiencing since our family formation with sincerity and spontaneity, self-organization, struggle and confrontation in the face of oppression, corruption, favoritism, interference, private appropriation of the public space, discrimination, environmental degradation, etc.; therefore, there are no formulas and models, because the involve(develop)ment is unique in each time-territory, it is experienced, lived amid the everyday complexities, temporalities, and territorialities.

Acknowledgements We are grateful for the financing granted by the National Council for Scientific and Technological Development, Brazil (CNPq)—Research Productivity Grant and Universal Tender (14/2013), *FundaçãoAraucária*—Basic and Applied Research Tender (24/2012) and *NAPI Alimento e Território*, as well as the Graduate Support Program of the Coordination for the Improvement of Higher Education Personnel, Brazil (PROAP/CAPES)—2020.

References

Bagnasco, A. (1977). *Tre italie. La problematica territoriale dello sviluppo italiano*. Bologna,Il Mulino;

Bagnasco, A. (1978). Problematiche dello sviluppo e articolazione dell'analisi: un paradigma per l'analisi territoriale. In A. Bagnasco, M. Messori, & C. Trigilia (Eds.), *Le problematiche dello sviluppo italiano* (pp. 205–251). Milão: Feltrinelli.

Bagnasco, A. (1988). *La costruzione sociale del mercato*. Bologna, Il Mulino

Bartra, A. (2014a). Campesinos del tercer milenio: Aproximaciones a una quimera. *Revista ALASRU, 10*, 17–43.

Bartra, A. (2014b). Campesindios: Ethos, clase, predadores, paradigma. Aproximaciones a una quimera. In F. Hidalgo, F. Houtart, P. Lizárraga (Eds.), *Agriculturas campesinas en Latinoamérica—Propuestas y desafíos*. Quito, Editorial IAEN, pp. 269–276.

Becattini, G. (2000[1989]a). Il distretto industriale marshalliano come concetto socio-economico. In G. Becattini (Ed.), *Il distretto industriale* (pp. 57–78). Torino: Rosenberg & Sellier.

Becattini, G. (2000). *Il distretto industriale*. Torino, Rosenberg and Sellier.

Becattini, G. (2009). *Ritorno al territorio*. Bologna, Il Mulino.

Becattini, B. (2009 [2002]). Costruire il territorio. In: Becattini, G. (Ed.) *Ritorno al territorio*, pp. 265–280. Bologna, Il Mulino.
Becattini, G. (2009). *Ritorno al territorio*. Bologna: Il Mulino.
Becattini, B. (2015). *La coscienza dei luoghi. Il territorio come soggetto corale*. Roma, Donzelli
Becattini, G., & Magnaghi A. (2015). Coscienza di classe e coscienza di luogo. In G. Becattini (Ed.), *La coscienza dei luoghi*, pp. 115–222. Roma, Donzelli.
Belliggiano, A., & De Rubertis S. (2012). Le filiere corte agroalimentari nei processi di sviluppo locale, XXIV Convegno di Sinergie, pp. 513–524.
Braudel, F. (1992 [1969]). *Escritos sobre a História*. Editora Perspectiva, São Paulo;
Braudel, F. (1981). *[1977])*. Bologna, Il Molino: La dinamica del capitalismo.
Brunetta, G. (1997). Reti di trasporto e asseti territoriali. Politiche di integrazione tra reti e strategie di efficienza allocativa. In R. Camagni & R. Capello (Eds.), *Strategie di competitività territoriale: il paradigma a rete*, pp. 119–148. Torino, SEAT.
Calvino, Í. (2008 [1972]). *As cidades invisíveis*. São Paulo, Companhia das Letras.
Camagni, R. (1990). Strutture urbane gerarchiche e reticolare: Verso una teorizazzione. In F. Curti, & L. Diappi (Eds.), *Gerarchie e reti di città: tendenze e politiche*. Milão, Franco Angeli.
Camagni, R. (1993). Le reti di città in Lombardia: introduzione e sintesi della ricerca. In R. Camagni, & G. De Blasio (Eds.), Le reti di città. Teoria, politiche e analisi nell'area padana, pp. 21–52. Milão, Franco Angeli.
Camagni, R. (1997). Luoghi e reti nelle politiche di competitività territoriale. In R. Camagni & R. Capello (Eds.), *Strategie di competitività territoriale: il paradigma a rete SEAT*, pp. 167–179. Torino.
Camagni, R., & Salone C. (1993). Elementi per una teorizzazione delle reti di città. In R. Camagni, & G. De Blasio (Ed.), *Le reti di città. Teoria, politiche e analisi nell'area padana*, pp. 53–67. Milão, Franco Angeli.
Chaffotte, L. & Chiffoleau, Y. (2007). Vente directe et circuits courts: évaluations, définitions et typologie, *Les Cahiers de L'Oservatoire CROC, 1*, 1–8.
Chávez, A., & Salcido, G. (2014). *Gobernanza de los sistemas agroalimentarios localizados: el caso de los productores rurales de nopal en Tlalnepantla, Morelos,* México: Universidad Nacional Autónoma de México.
Chamorro, G. (1998). A espiritualidade guarani: Uma teologia ameríndia da palavra. São Leopoldo, SINODAL.
Coraggio, C. (2004 [2000]). Desarrollo local y municipios participativos. In: J. Coraggio (Ed). *Descentralizar: Barajar y dar de nuevo. La participación en juego*, pp. 159–172. Quito, FLACSO Equador.
Corsi, A., et al. (2018). Multidisciplinary approaches to alternative food networks. In: A. Corsi et al. (Ed.), *Alternative food networks: An interdisciplinary assessment*, pp. 9–46. Cham, Switzerland, Palgrave Macmillan.
Dansero, E. (2008). Geografia e cooperazione allo sviluppo. Prospettive di ricerca. In E. Bignante, E. Dansero, C. Scarpocchi (Eds.), *Geografia e cooperazione allo sviluppo. Temi e prospettive per un approccio territoriale*, pp. 9–26. Milão, Franco Angeli.
Dansero, E., & Puttilli, M. (2014). Multiple territorialities of alternative food networks: six cases from Piedmont, Italy. *Local Environment, 19*(6), 626–643.
Deda, P. (1997). Reti di città e città in rete: politiche, sinergie progettuali ed alleanze strategiche. In R. Camagni & R. Capello (Eds.), *Strategie di competitivitàterritoriale: il paradigma a rete* (pp. 97–117). Torino: SEAT.
Dematteis, G. (1964). Alcuni relazioni tra l'ambito territoriale dei rapporti sociali e i caratteri della casa rurale, Atti 19. Congresso Geografico Italiano, III, pp. 239–253;
Dematteis, G. (1981 [1979]). Il "terreno" come lotta di classe: la "scoperta" del territorio nel 1968–69. In F. Canigiani, M. Carazzi, & E. Grottanelli (Eds.), *L'inchiesta sul terreno in geografia–geografia democratica* (pp. 135–144). Torino: Giappichelli.
Dematteis, G. (1985). *Le metafore della terra. La geografia umana tra mito e scienza*. Milano, Feltrinelli.

Dematteis, G. (1995). *Progetto implicito. Il contributo della geografia umana alle scienze del territorio.* Milano, Angeli.
Dematteis, G. (2001). Per uma geografia della territorialità attiva e dei valori territoriali. In P. Bonora (Ed.), *Slot, quaderno 1* (pp. 11–30). Bologna: Baskerville.
Dematteis G., & Governa F. (2003). Ha ancora senso parlare di identità territoriale? Atti dei Convegni Lincei, 194, Roma, Accademia Nazionale dei Lincei, pp. 264–281
Dematteis, G. (1994). Possibilità e limiti dello sviluppo locale. *Sviluppo locale, I* (1)10–30.
Dussel, E. (1995 [1979]). *Introducción a la filosofía de la liberación. Bogotá.* Editorial Nueva América.
Fals, B. O. (1981). La ciencia y el pueblo. In F. Grossi, V. Gianotten, T. Wit (Eds.), *Investigación participativa y praxis rural*, pp. 19–47. Lima, Mosca Azul.
Fals B. O. (2012 [1971]). La antiélite y su papel en el cambio social. In Herrera Farfán, N., López Guzmán, L. (Eds.), *Ciencia, compromiso y cambio social. Textos de Orlando Fals Borda*, pp. 109–117.. Buenos Aires, El Colectivo – Lanzas y Letras – Extensión Libros.
Fals B. O. (2012 [1972]). Reflexiones sobre la aplicación del método de estudio-acción en Colombia. In N. Herrera Farfán, & L. López Guzmán (Eds.), *Ciencia, compromiso y cambio social. Textos de Orlando Fals Borda*, pp. 241–252. Buenos Aires: El Colectivo – Lanzas y Letras – Extensión Libros.
Fals B. O. (2012 [1981]). Es posible una sociología de la liberación? In N. Herrera Farfán, L. López Guzmán (Eds.), *Ciencia, compromiso y cambio social. Textos de Orlando Fals Borda*, pp. 149–154. Buenos Aires: El Colectivo – Lanzas y Letras – Extensión Libros.
Fals B. O. (2012 [1986]). Primera lección: Saber interactuar y organizarse. In: N. Herrera Farfán & L. López Guzmán (Eds.), Ciencia, compromiso y cambio social. Textos de Orlando Fals Borda, pp. 123–141. Buenos Aires: El Colectivo – Lanzas y Letras – Extensión Libros.
Fals B. O. (2012 [1997]). La ciencia y el pueblo: nuevas reflexiones sobre la investigación-acción (participativa). In: N. Herrera Farfán, L. López Guzmán (Eds.), *Ciencia, compromiso y cambio social.Textos de Orlando Fals Borda*, pp. 301–319. Buenos Aires: Ed. El Colectivo – Lanzas y Letras – Extensión Libros.
Fals B. O. (2012a [2003]). La investigación participativa y la geografía. In N. Herrera Farfán, L. López Guzmán (Eds.), *Ciencia, compromiso y cambiosocial. Textos de Orlando Fals Borda*, pp. 327–331. Buenos Aires: Ed. El Colectivo – Lanzas y Letras – Extensión Libros.
Fals B. O. (2012b [2003]). Ciencias sociales, integración y endogénesis. In N. Herrera Farfán, L. López Guzmán (Eds.), *Ciencia, compromiso y cambio social. Textos de Orlando Fals Borda*, pp. 81–90.. Buenos Aires: Ed. El Colectivo – Lanzas y Letras – Extensión Libros.
Fals B. O., & Mora-Osejo L. E. (2004). La superación del eurocentrismo—enriquecimiento del saber sistémico y endógeno sobre nuestro contexto tropical. *POLIS—Revista Latinoamericana, 7*, 1–6.
Forno, F., & Maurano, S. (2016). Cibo, sostenibilità e territorio—dai sistemi di approvvigionamento alternativi ai food policy councils. *Rivista Geografica Italiana, 123*, 1–20.
Feyerabend, P. (2010 [1987]). *Adeus à razão.* São Paulo, (Eds.) Unesp.
Freire, P. (2011 [1996]). *Pedagogia da autonomia. Saberes necessários à prática educativa.* São Paulo, Paz e Terra.
Godoy, D. (2016). Territorios campesinos y agroindustria: un análisis de las transformaciones territoriales desde la economía de la proximidad. *Eutopía, 10*, 41–55.
Governa, F. (2001). Il territorio come soggetto collettivo? Comunità, attori, territorialità. In P. Bonora (Ed.), *Slot, quaderno 1* (pp. 31–46). Bologna: Baskerville.
Gottmann, J. (1973). *The significance of territory*. Charlottesville: University Press of Virginia.
Gramsci, A. (1975). Quaderni del carcere, vol. I (1929–1932). Turim, Einaudi.
Grignoli, D. (2019). La co-creazione dello sviluppo locale. In D. Grignoli, N. Bortoletto (Ed.), *Dal locale al globale e ritorno: Nuovi paradigmi e nuovi modelli di azione*, pp. 25–38. Milano, Franco Angeli.
Harguindeguy, L. C. (2014). *Economía solidaria: Local y diversa. Talaxcala.* México: El Colegio de Tlaxcala.

Harvey, D. (1982). O trabalho, o capital e o conflito de classes em torno do ambiente construído nas sociedades capitalistas abancadas, pp. 6–35. Espaço e Debates, 2, 6, São Paulo.
Hernández, C. A. (2008). De la dialéctica a la trialéctica del espacio: Proximaciones al pensamiento de Milton Santos y Edward Soja. In Mendoza C. (Ed.), *Tras las huellas de Milton Santos. Una mirada a la geografía humana contemporánea*, pp. 84–97. Barcelona,Anthropos/UAMI
Hidalgo, F., & Fernández, Á. (2012). Contrahegemonia y buen vivir. Quito: Universidad Central del Ecuador y Universidad del Zulia (Venezuela).
Indovina F., & Calabi, D. (1974). Sull'uso capitalistico del territorio. In: G. Lusso (Ed.), Economia e territorio. Milano, Angeli, pp. 205–222.
Kosik, K. (1976 [1963]). *Dialética do concreto.* Rio de Janeiro, Paz e Terra.
Kropotkin, P. (1982 [1885]). Lo que la Geografia debe ser. In: J. Mendoza, J. Jiménez & N. Cantero (Ed.), *El pensamiento geográfico—estudio interpretativo y antologia de textos*, pp. 227–240. Madri: Alianza Editorial.
Lizárraga, P., & Vacaflores, C. (2014). Lo agrario en la reconstitución del sujeto indígena originario campesino. In: F. Hidalgo, F. Houtart & P. Lizárraga (Eds.), *Agriculturas campesinas en Latinoamérica—propuestas y desafios*, pp. 277–296. Quito, Editorial IAEN.
Magnaghi, A. (1976). Il territorio nella crisi. *Quaderni Del Territorio, 1*(1), 15–29.
Magnaghi, A. (2000). *Il progetto locale.* Torino, Bollati Boringhieri.
Magnaghi, A. (2003). La rappresentazione identitaria del patrimonio territoriale. In G. Dematteis, & F. Ferlaino (Eds.), *Il mondo e i luoghi: geografie delle identità e del cambiamento*, pp. 13–20. Torino: IRES/SGI.
Magnaghi, A. (2013). Riterritorializzare il mondo. *Scienze Del Territorio, 1*, 47–58.
Magnaghi, A. (2015). La lunga marcia del ritorno al territorio. In G. Becattini (Ed.), *La coscienza dei luoghi*, pp. VII-XVI. Roma, Donzelli.
Mansilla Q. P., Quintero W. J., & Moreira-Muñoz A. (2019). Geografía de las ausencias, colonialidad del estar y el territorio como sustantivo crítico en las epistemologías del Sur. *Utopía y praxis latinoamericana, 24*(86), 148–161.
Mannheim, K. (1952). On the interpretation of weltanschauung. In K. Mannheim, (Eds.) *Essays on the sociology of knowledge.* (pp. 33–83) London: Routledge and Kegan Paul.
Melin M., Mansilla Q. P., Royo M. (2016). *Descolonizando el Mapa del Wallmapu, construyendo cartografía cultural en Territorio Mapuche.* Temuco, Chile, Pu Lof Editores.
Pecqueur and Zimmermann (2002). *Les fondementsd'une économie de proximités*, pp. 1–23. Marseille, GREQAM, DT 02A26.
Quaini, M. (2004). L'elogio dei luoghi e la voglia di pre-moderno. *Riflessioni in Margine a Un Manuale Curato Da Alberto Magnaghi, Rivista Geografica Italiana, 111*, 341–355.
Quaini, M. (2006). *L'ombra del paesaggio. Orizzonti di un'utopia conviviale.* Reggio Emilia, Diabasis.
Quaini, M. (2010). *Dalla coscienza di classe alla "coscienza di luogo" ovvero "de la lutte des classes à la lutte des places*, pp. 1–13. Declinazioni del concetto di luogo e di paesaggio. Treviso, Fondazione Benetton.
Quijano, A. (2000). El fantasma del desarrollo en América Latina. *Revista Venezolana de Economía y Ciencias Sociales, 6* (2)73–90
Quintero, W. J. (2011). Wopukaru Jatumi Wataawai: el camino hacia nuestro propio saber. *Utopía y Praxis Lationoamericana, 16*(54), 93–116.
Raffestin, C. (1977). Paysage et territorialitè. *Cahiers De Géographie Du Québec, 21*(53–54), 123–134.
Raffestin, C. (1993 [1980]). *Por uma geografia do Poder*. São Paulo, Ática.
Raffestin, C. (1984). Territorializzazione, deterritorializzazione, riterritorializzazione e informazione. In A. Turco (Ed.), *Regione e regionalizzazione*, pp. 69–82. Milano, Angeli.
Raffestin, C. (1986). Punti di riferimento per una teoria della territorialità umana. In C. Copeta (Ed.), *Esistere ed abitare. Prospettive umanistiche nella geografia francofona*, pp. 75–89. Milano, Angeli.

Raffestin, C. (1987). Reperès pour une théorie de la territorialitè humaine. *Cahier/groupe Reseaux, 7*(3), 2–22.
Raffestin, C. (2005). *Dalla nostalgia del territorio al desiderio di paesaggio. Elementi per una teoria del paesaggio.* Firenze, Alinea.
Renting, H., Marsden, T., & Banks, J. (2003). Understanding alternative food networks: Exploring role of short food supply chains in rural development. *Environment and Planning A, 35*, 393–411.
Reyes, L.A. (2009). *El pensamiento indígena en América.* Buenos Aires, Biblos.
Rullani, E. (1997). Più locale e più globale: Verso una economia postfordista del territorio. In A. Bramanti & M. Maggioni (Eds.), La dinamica dei sistemi produttivi territoriali: Teorie, tecniche, politiche, pp. 85–111. Milano, Franco Angeli.
Rullani, E. (2003). Complessità sociale e intelligenza localizzata. In G. Garofoli, (Ed.), *Impresa e territorio*, pp. 85–130. Bologna, Il Mulino.
Rullani, E. (2005). Dai distretti alla distrettualizzazione: le forze che fanno (e disfano) il sistema territoriale. In N. Bellanca, M. Dardi, T. Raffaelli (Ed.), *Economia senza gabbie*, pp. 111–169. Bologna, Il Mulino
Rullani, E. (2009). La città al tempo delle reti. In *Le frontiere della Geografia*, pp. 145–172. Novara-Torino: UTET.
Rullani, E., Micelli, S., & DiMaria, E. (2000). Tra città reale e città virtuale: il territorio nel post-fordismo. In E. Rullani, S. Micelli, & E. Di Maria (Eds.), *Città e cultura nell'economia delle reti* (pp. 13–56). Bologna: Mulino.
Rullani, E. (2009). La città al tempo delle reti. *Le frontiere della Geografia.* 145–172 Novara-Torino: UTET.
Sánchez, A. (2014c). Dilemas teóricos y metodológicos de la sociología rural en América Latina. *Revista ALASRU, 10*, 359–380.
Sántiz-Gómez, A. (2015). Ts'Umbal activo en las transformaciones rurales de Oxchuc. *Chiapas, Revista LiminaR – Estudios Sociales y Humanísticos, 13*(2), 122–134.
Sántiz-Gómez A., & Parra, V. M. (2017). *Lavisóntseltal de la vida en el desarrollo alternativo de Oxchuc, Chiapas.* In: García García, A. (Ed.), *Extractivismo y neoextractivismo en el sur de México: múltiples miradas*, pp. 317–350. Texcoco, México, Universidad Autónoma Chapingo.
Sanz, J. (2014). Sistemas agroalimentarios locales y multifuncionalidad. Un enfoque de investigación en alimentos, ciencias sociales y territorio. In M. Rivera (Ed.), *El desarrollo hoy hacia la construción de nuevos paradigmas*, pp. 87–103. México: UNAM – Instituto de Investigaciones Económicas.
Santos, M. (1994). O retorno do território. In: M. Santos, et al. (Eds.), *Território: globalização e fragmentação*, pp. 15–20. São Paulo,Hucitec/ANPUR.
Saquet, M. (2003 [2001]). *Os tempos e os territórios da colonização italiana.* Porto Alegre, EST Edições.
Saquet, M. (2009). Por uma abordagem territorial. In M. Saquet & E. Sposito (Eds.), *Territórios e territorialidades: teorias, processos e conflitos* (pp. 73–94). São Paulo: Expressão Popular.
Saquet, M. (2013). Por uma abordagem territorial: continuando a reflexão. In M. Saquet (Ed.), *Estudos territoriais na ciência geográfica* (pp. 47–74). São Paulo: Outras Expressões.
Saquet, M. (2014). Territorialidades, relações campo-cidade e ruralidades em processos de transformação territorial e autonomia. *Revista Campo-Território, 9*(18), 1–30.
Saquet, M. (2015[2011]). *Por uma Geografia das territorialidades e das temporalidades: uma concepção multidimensional voltada para a cooperação e para o desenvolvimento territorial.* Rio de Janeiro: Editora Consequência.
Saquet, M. (2016a). Territory, geographicalindicationand territorial development. *Desenvolvimento Regional Em Debate, 6*, 4–21.
Saquet, M. (2016b). Territorios rurales y perspectivas de desarrollo territorial con autonomía: la agricultura campesina (agro)ecológica. *Eutopía, 10*, 57–76.
Saquet, M. (2017). *Consciência de classe e de lugar, práxis e desenvolvimento territorial.* Rio de Janeiro: Consequência.

Saquet, M. (2021 [2017]). *Conciencia de clase y de lugar, praxis y desarrollo territorial.* Buenos Aires, CLACSO.
Saquet, M. (2018). A descoberta do território e outras premissas do desenvolvimento territorial. *Revista Brasileira De Estudos Urbanos e Regionais, 10*(3), 479–505.
Saquet, M. (2020[2007]). *Approaches and conceptions of territory.* Curitiba: Brazil Publishing.
Saquet, M. (2021). *Popular knowledge, territorial praxis and counter-hegemony.* Curitiba, BrazilPublishing.
Saquet, M. (2022). *Singularidades: um manifesto a favor da ciência territorial popular feita na práxis descolonial e contra-hegemônica.* Rio de Janeiro, Editora Consequência.
Saquet, M., & Alves, A. (2015). Desarrollo territorial heterocentrado y autocentrado: diferentes formas de movilizar saberes y redes en Brasil. *Revista Textual, 65*, 11–34.
Saquet, M., Dansero E., Candiotto L. (Eds.) (2012). *Geografia da e para acooperação ao desenvolvimento territorial: experiências brasileiras e italianas.* São Paulo, Outras Expressões;
Scoppetta, C. (2009). *Immaginare la metropoli della transizione. La città come living machine.* Roma, Campisano;
Storper, M. (2003). *Le economie locali come beni relazionali.* In: Garofoli, G. (Ed.), *Impresa e territorio*, pp. 169–208. Bologna, Il Mulino.
Turco, A. (1988). Verso uma teoria geografica della complessità. Milano, Unicopli.
Turco, A. (2010). *Configurazioni della territorialità.* Milão Franco Angeli
Vazquez, A. (1990[1977]). *Filosofia da Praxis.* Rio de Janeiro: Paz e Terra.

Marcos Aurelio Saquet Doctor in Geography from Unesp—PresidentePrudente (São Paulo, Brazil) and PhD from the Polytechnic and Universitàdi Torino—Italy. He is the author of different books, including "Enfoques y concepciones de territorio", 2019, Bogotá, Editorial Universidad Distrital F. J. de Calcas; "Popular knowledge, territorial praxis and conter-hegemony", Curitiba, Brazil, 2021, BrazilPublishing; "Approaches and conceptions of territory", Curitiba, Brazil, 2020, BrazilPublishing. He is an evaluator for different funding agencies, such as CNPq, FAPESP, CONACYT—Mexico, and CNR—Italy. He has been a visiting professor at different universities (face-to-face or remote classes) in Colombia, Argentina, Costa Rica, Italy, Uruguay, and Mexico.

Index

M. Mishra et al. (eds.), *Contemporary Geopolitical Alignment in the Global South*, https://doi.org/10.1007/978-3-031-81977-3

MIX
Papier aus verantwortungsvollen Quellen
Paper from responsible sources
FSC® C105338

If you have any concerns about our products,
you can contact us on
ProductSafety@springernature.com

In case Publisher is established outside the EU,
the EU authorized representative is:
Springer Nature Customer Service Center GmbH
Europaplatz 3, 69115 Heidelberg, Germany

Printed by Libri Plureos GmbH
in Hamburg, Germany